DISCOVERING CALCULUS WITH DERIVE®

Second Edition

JERRY JOHNSON
UNIVERSITY OF NEVADA, RENO

BENNY EVANS
OKLAHOMA STATE UNIVERSITY

JOHN WILEY & SONS, INC.
New York Chichester Brisbane Toronto Singapore

ISBN 0-471-00972-5

Printed in the United States of America

10 9 8 7 6 5 4 3 2 1

Printed and bound by Malloy Lithographing, Inc.

Table of Contents

PREFACE

This second edition of **Discovering Calculus with *DERIVE***, like the first, was written as an enrichment supplement to an otherwise traditional calculus course. Its purpose is to help students use the *DERIVE*® program as a tool to explore calculus beyond the level of rote calculations and template exercises by providing problems that are different from those one would normally expect to do with nothing but pencil and paper. We have both taught calculus using *DERIVE* and have found it a wonderful help both in the classroom and as a laboratory tool.

The second edition is considerably different from the first. Instead of including an extensive and comprehensive set of exercises as we did in the first edition, we concentrated on multi-step structured laboratory assignments that we think offer students a better learning experience. Most the exercises ask students for explanations.

Key Features of the Second Edition

- No prior knowledge of *DERIVE* is required. Novices learn the basics quickly in the Introduction which includes a tear-out Quick Reference section. Appendix I provides more detailed explanations of features of the *DERIVE* program.
- There are more **Solved Problems** with more explanation than the first edition.
- Each chapter heading contains a list of calculus topics that are used and a list of new *DERIVE* features that are covered in the chapter.
- The first use of a new *DERIVE* topic in a **Solved Problem** is signaled by the **New *DERIVE* Lesson** heading, and a full explanation is given.
- The problems are presented as tear-out laboratory exercises with space for students to show their work.
- To ensure that students have gained the necessary skills, **Practice Problems**, complete with answers, precede each set of laboratory exercises.

- *DERIVE* code for creating graphics and automating some of the calculations is provided in Appendix II.
- An index of **Solved Problems** and **Laboratory Exercises** is provided.
- Most of the Laboratory Exercises have been class-tested by the authors and others.

Structure of the Book

The chapters roughly follow the flow of the fifth edition of **Calculus with Analytic Geometry** by Howard Anton, and the seventh edition of **Calculus One and Several Variables** by S. L. Salas, Einar Hille and revised by Garret J. Etgen, and the structure is similar to that of most traditional calculus books. Each chapter contains several **Solved Problems** that are followed by **Practice Problems** and **Laboratory Exercises**.

The **Solved Problems** are examples that provide a context for *DERIVE* instructions that students are likely to encounter in the exercises, but they contain useful mathematics hints and lessons as well.

The **Practice Problems** are brief exercises designed to build a student's confidence with *DERIVE*. Answers are provided.

The **Laboratory Exercises** are sets of problems that students should be able to do upon completion of the appropriate Solved Problem. They are sufficiently complex that solving them without assistance from a computer or graphing calculator is usually not practical. Each is carefully designed to teach students important mathematical ideas, and most of them invite the students to discuss their observations and findings.

DERIVE is very impressive, but our central purpose is to teach *mathematics*, not to show off hardware and software. We have tried to include enough *DERIVE* instructions and suggestions in the Introduction, the **Solved Problems**, and Appendix I so the reader will not have to spend a great deal of time referring to the *DERIVE* User Manual, but this is not intended as a substitute for it.

The *DERIVE*® Program

DERIVE is a comprehensive computer algebra system that requires only a 512K PC-compatible. There is a version that runs on HP "palm-top" computers and to take advantage of extended memory there is the more expensive product *DERIVE XM*. As of the printing of this book, there is no Macintosh version, but *DERIVE* will run under MicroSoft Windows as a full-screen DOS application. A Windows version of *DERIVE* is in development.

Any sophisticated software takes some practice and experience to master, but we have used *DERIVE* in our courses almost since its first release and are convinced that one of its strongest educational points is that it is easy for students to learn and to use. It is also powerful enough for professional applications in mathematics, science, and engineering.

This manual was written using *DERIVE* version 2.6. If you are using an earlier version, you will notice some differences which we have tried to mention on the rare occasions where they might cause confusion. The most conspicuous new feature in versions 2.5 and higher is in the **Plot** command. (See Section 5 in Appendix I.) Also, the arrow keys may now be used to move the cursor on the **Author** line as is explained in Section 2 in Appendix I.

Version 3 of *DERIVE* is scheduled for release in the Fall of 1994, and it includes many new features that make it even more powerful and easy to use. Among other things, the plotting features are much improved, and implicit plotting is implemented.

Suggestions for Incorporating *DERIVE*: What Has Worked for Us

We normally spend a period with our classes at the computer laboratory in the first week or two of the semester to introduce them to *DERIVE*. Our experience is that students quickly become comfortable with *DERIVE* and that only minimal help is needed later.

After this initial session, we may return to the lab with the class two or three more times during the semester. It is important to try to integrate the lab experience with material that is currently being covered in the course. We often begin such a lab session by asking students to work through a Solved Problem and later give them a quiz that consists of a similar exercise. Grades in the lab are important. Even the best students appreciate credit for their work, and there are always students who will remain passive in the lab without the incentive of a grade.

We give outside lab assignments from time to time. The frequency may vary from five to ten assignments a semester. We encourage students to work together on the assignments, but we insist that they turn in their own printouts and write up the results in their own words.

We have found that some classroom discussion of an assigned exercise is desirable, including a few words about new *DERIVE* commands. However, the **Solved Problems** and Appendix I should contain the essentials.

A Word to Students

DERIVE has essentially automated most of the standard algebra calculations you normally encounter, just as scientific calculators have done with arithmetic. It will simplify complicated expressions, solve equations, and draw graphs. It will also do most of the standard calculations that arise in calculus such as finding derivatives and integrals. But that doesn't mean calculus and algebra are obsolete or unimportant. Even though calculators will do arithmetic, we still have to know what questions to ask, understand what the answers mean, and realize when an obvious error has been made. In the same way, we still have to understand the definitions, concepts, and processes that are involved in algebra and calculus so we will know what to tell *DERIVE* to do, what its answers mean, and be able to detect errors. *DERIVE* only does the calculations; you must still do the thinking.

Always view any computer or calculator output critically. Be alert for answers that seem strange. You might have hit the wrong key, entered the wrong data or made some other mistake. It is even possible that the program has a bug! If a problem asks for the cost of materials to make a shoe box and you get $123.28 or −$0.55, you should suspect something is wrong!

Before you turn on the computer, work through as much of an assigned problem as you can with pencil and paper, taking note of exactly where you think the computer will be required and for what purpose. You may be surprised at how little time you will actually have to spend in front of the machine if you follow this advice.

Clear communication is *at least* as important in mathematics as in other fields. You should always write your answers neatly in complete, logical sentences. Before you turn in an assigned Laboratory Exercise, re-read what you have written and ask yourself if it really makes sense.

Benny Evans and Jerry Johnson

September 1994

Key for Use with
Calculus One and Several Variables, 7e by S.L. Salas and Einar Hille

Salas and Hille		Johnson and Evans	
Chapter 1:	Introduction	*Chapter 1:*	*Functions and Graphs*
Chapter 2:	Limits and Continuity	*Chapter 2:*	*Limits*
Chapter 3:	Differentiation	*Chapter 3:*	*Differentiation*
Chapter 4:	The Mean Value Theorem and *Applications*	*Chapter 4:*	*Applications of the Derivative*
Chapter 5:	Integration	*Chapter 5:*	*Riemann Sums and Integration*
Chapter 6:	Some Applications of the Integral	*Chapter 6:*	*Applications of the Integral*
Chapter 7:	The Transcendental Functions	*Chapter 7:*	*Logarithmic and Exponential Functions*
		Chapter 8:	*Hyperbolic and Inverse Trigonometric Functions*
Chapter 8:	Techniques of Integration	*Chapter 9:*	*Numerical Integration*
Chapter 9:	Conic Sections; Polar *Coordinates; Parametric Equations*	*Chapter 12:*	*Polar Coordinates and* Parametric Equations
Chapter 10:	Sequences, Indeterminate Forms, Improper Integrals	*Chapter 10:*	*Improper Integrals*
Chapter 11:	Infinite Series	*Chapter 11:*	*Infinite Series*
Chapter 12:	Vectors	*Chapter 13:*	*Vectors and Vector Valued Functions*
Chapter 13:	Vector Calculus	*Chapter 13:*	*Vectors and Vector Valued Functions*
Chapter 14:	Functions of Several Variables	*Chapter 14:*	*Partial Derivatives*
Chapter 15:	Gradients; Extreme values; Differentials	*Chapter 14:*	*Partial Derivatives*
Chapter 16:	Double and Triple Integrals	*Chapter 15:*	*Double Integrals and Line Integrals*
Chapter 17:	Line Integrals and Surface *Integrals*	*Chapter 15:*	*Double Integrals and Line Integrals*
Chapter 18:	Elementary Differential Equations	*Chapter 16:*	*Differential Equations*

Key for Use with
The Fifth Edition of Calculus with Analytic Geometry by Howard Anton

Anton		Johnson and Evans	
Chapter 1:	Coordinates, Graphs, Lines	*Chapter 1:*	*Functions and Graphs*
Chapter 2:	Functions and Limits	*Chapter 1:* *Chapter 2:*	*Functions and Graphs* *Limits*
Chapter 3:	Differentiation	*Chapter 3:*	*Differentiation*
Chapter 4:	Applications of Differentiation	*Chapter 4:*	*Applications of the Derivative*
Chapter 5:	Integration	*Chapter 5:*	*Riemann Sums and Integration*
Chapter 6:	Applications of the Definite *Integral*	*Chapter 6:*	*Applications of the Integral*
Chapter 7:	Logarithmic and Exponential *Functions*	*Chapter 7:*	*Logarithmic and Exponential Functions*
Chapter 8:	Inverse Trigonometric and *Hyperbolic Functions*	*Chapter 8:*	*Hyperbolic and Inverse Trigonometric Functions*
Chapter 9:	Techniques of Integration	*Chapter 9:*	*Numerical Integration*
Chapter 10:	Improper Integrals; L'Hôpital's Rule	*Chapter 10:*	*Improper Integrals*
Chapter 11:	Infinite Series	*Chapter 11:*	*Infinite Series*
Chapter 12:	Topics in Analytic Geometry	*Chapter 12:*	*Polar Coordinates and Parametric Equations*
Chapter 13:	Polar Coordinates and Parametric *Equations*	*Chapter 12:*	*Polar Coordinates and Parametric Equations*
Chapter 14:	Three-dimensional Space; Vectors	*Chapter 13:*	*Vectors and Vector-Valued Functions*
Chapter 15:	Vector-Valued Functions	*Chapter 13:*	*Vectors and Vector-Valued Functions*
Chapter 16:	Partial Derivatives	*Chapter 15:*	*Partial Derivatives*
Chapter 17:	Multiple Integrals	*Chapter 15:*	*Double Integrals and Line Integrals*
Chapter 18:	Topics in Vector Calculus	*Chapter 15:*	*Double Integrals and Line Integrals*
Chapter 19:	Second-Order Differential Equations	Chapter 16:	Differential Equations

Introduction

Getting Started

One of the advantages of *DERIVE* is that it is easy to use. Nonetheless, if you are a beginner we suggest that you read this chapter and work through the tutorials. Even if you are an experienced user, we recommend that you look over the chapter to familiarize yourself with the way *DERIVE* instructions are presented in this book. Each of the features of *DERIVE* covered here will be introduced in the context of solving a problem. Additional information on each of these topics appears in Appendix I.

Conventions

Boxed instructions: When a key is to be depressed on the keyboard, it will be put in a box. For example, [F3] [Enter] means you are to press the function key F3 and then press the Enter key in sequence. Sometimes you may have to hold down two keys at once. In this case, both keys will appear in the same box with a space between them. For example, to type the Greek letter π, you hold down the key marked Alt while you press the letter p. We will indicate this by [Alt p]. Similarly, [Ctrl →] means you hold down the key marked Ctrl while you press the right arrow key.

Commands from the menu: When we want you to execute a command from *DERIVE's* menu, shown in Figure 0.1, we will give the command in boldface spelled and formatted as it appears in the menu.

```
COMMAND: Author Build Calculus Declare Expand Factor Help Jump soLve Manage
         Options Plot Quit Remove Simplify Transfer moVe Window approX
Enter option
                                        Free:100%           Derive Algebra
```

Figure 0.1: **The basic *DERIVE* menu**

For example, in response to the instruction "**Simplify**," you would press the letter [S] to give *DERIVE* the command to simplify an expression. To approximate an expression, the command is **approX**, and you would press the [X] key.

Sometimes a sequence of commands is necessary, in which case we will list them. For example, to set *DERIVE* so that it will report answers in decimal form, we would say "use the commands **Options Notation Decimal**." When you press [O] for **Options**, you will be presented with a new menu. One of the selections from this menu will be **Notation**. Press [N] and you will be presented with a third menu, one of whose items is **Decimal**. Press [D] [Enter] and you have completed the instruction.

After the first chapter, we will shorten some instructions. Early in the book if we want you to approximate expression number 7, we will say "highlight expression 7 and **approX** [Enter]" thus providing a complete set of instructions. As you become familiar with *DERIVE*, this will be shortened to **approX** expression 7. Brief instructions are easier to read and should cause no difficulty.

Typing expressions: When you must enter an expression into *DERIVE*, it will be put in "typewriter" style, and you type exactly the symbols you see except the period at the end. For example, to enter $\frac{x^2}{3}$ you must first issue the **Author** command and then type `x^2/3` followed by [Enter]. We will indicate this with "**Author** `x^2/3`."

Upper and lower case: In its default state *DERIVE* is case insensitive, but generally everything is displayed in lower case except function names. When we ask you to enter expressions, we will follow *DERIVE's* convention, but in fact you do not need to worry about case. For example, to enter the function $\sin(x + y)$ our instruction will be **Author** `SIN(x+y)`. You may type `SIN(x+y)`, `sin(x+y)`, `SIN(X+Y)`, or even `SiN(X+y)`. No matter which you use, *DERIVE* will display $\text{SIN}(x + y)$.

Pictures: The pictures you see in this book will often show exactly what you see on your screen, as in Figures 0.1 and 0.2, but sometimes we will save space by splitting the screen into several "windows." You do not have to do this, but if you wish to learn how, consult your *DERIVE* user's manual.

Tutorial I: Algebra and Arithmetic

DERIVE is driven by a menu at the bottom of the screen. You can select an item in two ways: (1) Type the capitalized letter in the name. For example, to select **soLve** press [L]. (2) Use the [Spacebar] to move through the menu and press [Enter] to select the highlighted item.

Entering an expression involves customary syntax: addition (+), subtraction (−), division (/), exponents, (^), and multiplication (∗). However, multiplication does not require a ∗; `2x` is the same as `2*x`.

If you need help during a session, press [H] for **Help**, and *DERIVE* will provide references to the *DERIVE* User Manual for the topic you choose.

Arithmetic: Use *DERIVE* to calculate the exact value of $\frac{1}{3}+\frac{1}{7}$.

From the menu select **Author**, type the expression `1/3+1/7` on the *author line*, and press [Enter] when you are done. You should see $\frac{1}{3}+\frac{1}{7}$ as displayed in expression 1 of Figure 0.2. To perform the addition press [S] for **Simplify**. The result appears as expression 2. This is typical of the way *DERIVE* handles input: it formats it nicely on the screen but does not perform an operation on the expression until you tell it to do so.

1: $\frac{1}{3}+\frac{1}{7}$

2: $\frac{10}{21}$

3: 0.476190

4: π

5: 3.1415926535897932384626433832795028841971693993751

6: 100!

7: 9332621544394415268169923885626670049071596826438162146859296389521759999322

8: $\pi^{0.47619}\left[\frac{1}{3}+\frac{1}{7}\right]$

9: 0.821332

Figure 0.2: **Arithmetic tutorial**

Decimal approximations: Obtain a decimal approximation to $\frac{1}{3}+\frac{1}{7}$.

With expression 2 highlighted, press [X] for **approX**, and the decimal approximation appears in expression 3 of Figure 0.2. We should note that if you are interested in the approximate value only, then the intermediate step of first obtaining the exact answer can be skipped. Just **Author** `1/3+1/7` and **approX**.

Setting precision: Obtain a 50-place decimal approximation of π.

First **Author** `pi` and [Enter]. In its default setting *DERIVE* provides six digit approximations as we saw in the last example. This can be changed to any number of digits you wish. We use **Options Precision** [Tab], change Digits:6 to Digits:`50`, and press [Enter]. (If you make a mistake paging through the menus, use [Esc] to back up.) Now use **approX**, and *DERIVE* will present you with the approximation in expression 5. Before proceeding, you should return *DERIVE* to its default setting using **Options Precision** [Tab] `6`.

Viewing longer expressions: Calculate 100!.

Author `100!` , press [Enter], and **Simplify**. The result is partially displayed in expression 7 of Figure 0.2, but it is too long to fit on the screen. To see the rest of it, use [Ctrl →] and [Ctrl ←].

Building expressions: Approximate $\pi^{0.476190}\left(\frac{1}{3}+\frac{1}{7}\right)$.

We could type this expression on the author line and **approX**, but we will offer two ways to do this that avoid unnecessary typing:

Referring to expressions by line numbers: You can refer to expressions by their line numbers. Notice that π appears in expression 4, 0.476190 is in expression 3, and $\frac{1}{3}+\frac{1}{7}$ is in expression 1. Thus we **Author** `#4^#3(#1)` and then **approX**. The result appears in expression 9 of Figure 0.2.

Using the [F3] key and the arrow keys: The other way to enter $\pi^{0.476190}\left(\frac{1}{3}+\frac{1}{7}\right)$ uses the arrow keys and the [F3] and [F4] keys. The up and down arrow keys, [↑] [↓], move the highlight among the expressions on the screen. First issue the **Author** command and then use the arrow keys to highlight expression 4. Now press the function key [F3], and π is brought to the author line. Next type `^`, use the arrow keys to move the highlight to expression 3, and press [F3] once more. Finally, highlight expression 1 and press [F4] which encloses expression 1 in parentheses before bringing it to the author line. Press [Enter] to complete the expression, and then **approX**.

Factoring expressions: Factor x^2+3x-4.

Author `x^2+3x-4` and [Enter]. Press [F] for **Factor** and [Enter] twice to accept the default factor options. The factored form, $(x-1)(x+4)$, appears as expression 11 of Figure 0.3.

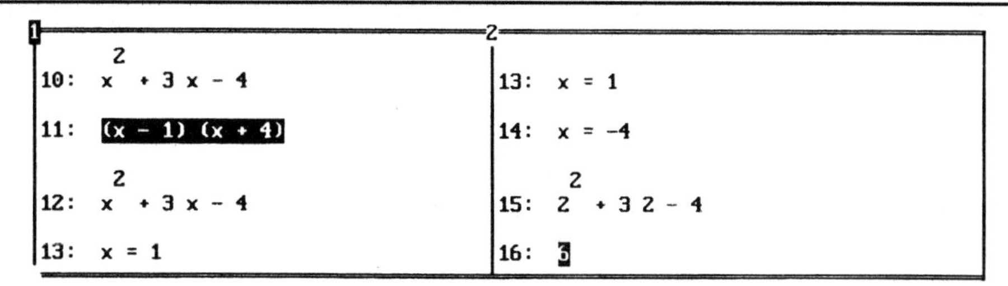

Figure 0.3: **Algebra tutorial**

Expanding expressions: Expand the product $(x-1)(x+4)$.

With expression 11 highlighted, choose the **Expand** command and Enter. Now expression 11 has been expanded to its original form and appears as expression 12.

Solving equations: Find the solutions of $x^2+3x-4=0$.

With expression 12 highlighted, press L for **soLve**. *DERIVE* lists the two solutions to the equation as expressions 13 and 14. When you ask *DERIVE* to **soLve** an expression that is not an equation as we did here, it attempts to find the zeros.

Substituting expressions: Evaluate x^2+3x-4 at $x=2$.

First use the arrow keys to highlight expression 12, and then use the commands **Manage Substitute** Enter. You will be prompted with MANAGE SUBSTITUTE value: x . Type 2 to tell *DERIVE* that you want to replace x by 2, and then Enter. We see $2^2+3\ 2-4$ in expression 15. **Simplify** to get the answer, 6, in expression 16.

Tutorial II: Plotting

Plotting a graph: Plot the graph of $\sin x$.

Author `SINx` and press P for **Plot**. You will be prompted for a choice: Beside Under Overlay. (Note: In versions of *DERIVE* earlier than 2.55 you are not offered this choice. The overlay option is automatic.) The **Beside** option opens a plot window beside your work, the **Under** option opens a plot window below your work, and the **Overlay** option uses the entire screen for graphing. We will usually use **Beside**; press B to select this option. Next you are prompted for a column number where *DERIVE* will split the screen. The default, 40, divides the screen in half. Press Enter to accept 40. After the plot window has been opened, press [P] for **Plot** once more, and the graph will appear as in Figure 0.4.

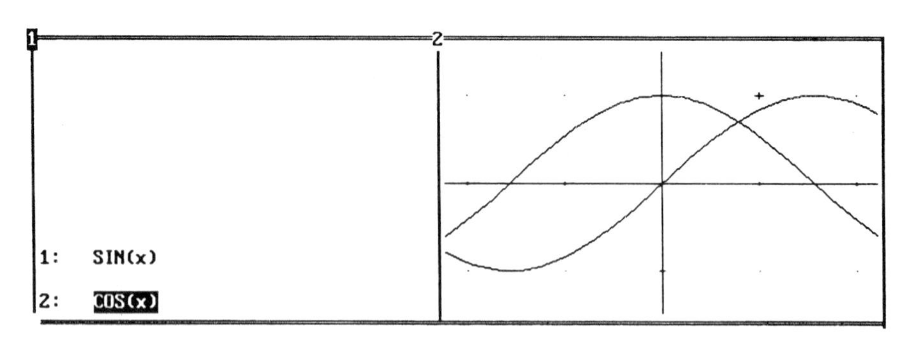

Figure 0.4: **The graphs of** $\sin x$ **and** $\cos x$

Moving between windows: When you have more than one window open, they are numbered in the upper left-hand corner. The highlighted number tells you which window you are in. From the algebra window **Plot** moves you to the plot window, and from there **Algebra** moves you back to the algebra window.

Multiple graphs: Include the graph of $\cos x$ in the picture with $\sin x$.

Use the **Algebra** command to return to the algebra window. **Author** COSx, **Plot** to return to the graphing window, and **Plot** again to view the graph. You will notice that *DERIVE* does not erase existing graphs when you ask it to draw new ones. If you want to get rid of old graphs, you must use the **Delete** command as explained later.

Zooming and centering: Zoom in to approximate the zero of $\cos x$ between 0 and π.

The correct answer is of course $\frac{\pi}{2}$, but we will approximate the answer graphically. In the plot window the arrow keys move the *graphing cross*, and its coordinates appear at the lower left-hand corner of the screen. [PageUp], [PageDown], [Ctrl →], and [Ctrl ←] move the cross in bigger jumps. Move the cross as close as you can to the first point to the right of the y-axis where the graph of $\cos x$ crosses the x-axis and **Center**. Now press [F9] to zoom in for a closer view. ([F10] zooms out.) Adjust the location of the graphing cross with the arrow keys, **Center**, and use [F9] to zoom in once more. In the lower left corner of the screen in Figure 0.5 we see Cross x:1.5722 y:0. Thus the approximate value of the zero is 1.5722. You may **Center** and zoom in further to improve accuracy. In Chapter 1 we will present a quick method for finding zeros that guarantees accuracy to any desired degree.

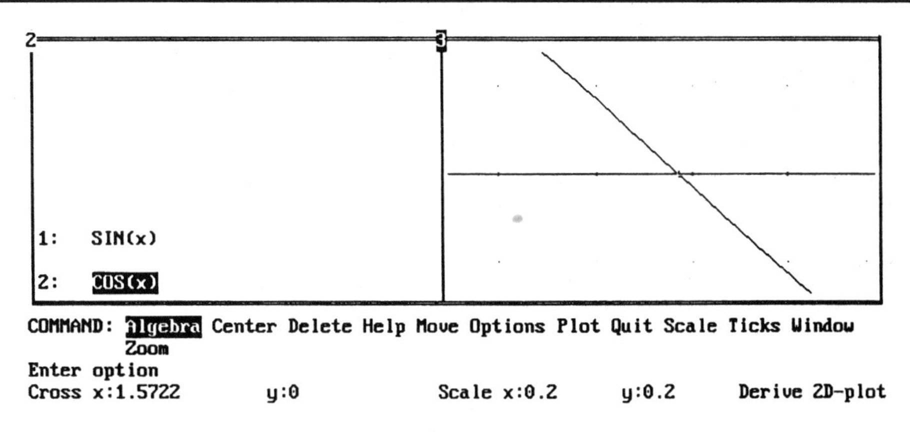

Figure 0.5: **Zooming in on a zero**

Setting the scale manually: Graph $\sin x$ and $\cos x$ with a scale of 2 on the x-axis and 0.5 on the y-axis.

From the plot window use [S] to issue the **Scale** command. You will be prompted with Scale x:1 y:1. Change x:1 to x:2 and use the [Tab] key to move to the y field. Change this to y:`0.5` and [Enter].

Cleaning up and closing the window: To get rid of old graphs, use **Delete All**. To close the plot window, use **Window Close**. Respond with `Y` for yes when prompted with Close Window 2 (y/n)?

Tutorial III: Functions, Constants, and Editing

Defining functions: Let $f(x) = x^2 + 1$. Calculate $f(3)$ and $f(t+5)$.

To define f, we **Author** `F(x):=x^2+1`. (Note: The colon is essential! *DERIVE* uses "=" for an equation and ":=" to indicate a definition.) *DERIVE* will now interpret f as this function until we "undefine it" by **Author**ing `f:=` . To get $f(3)$, **Author** `F(3)` and **Simplify**. The answer appears in expression 3 of Figure 0.6. To get $f(t+5)$, **Author** `F(t+5)` and **Simplify**. The result is in expression 5.

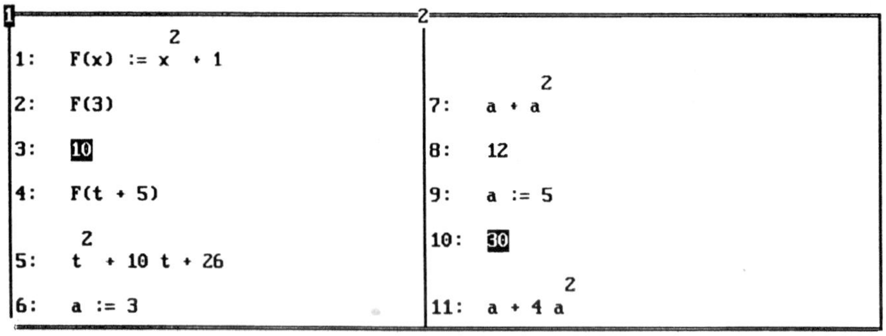

```
1:   F(x) := x^2 + 1            7:   a + a^2
2:   F(3)                       8:   12
3:   10                         9:   a := 5
4:   F(t + 5)                   10:  30
5:   t^2 + 10 t + 26            11:  a + 4 a^2
6:   a := 3
```

Figure 0.6: **Functions and constants**

Clearing function definitions: If you no longer wish f to be a function, **Author** `f:=` and [Enter]. This will instruct *DERIVE* to forget the definition and treat f as a variable.

Defining constants: Calculate $a + a^2$ in case $a = 3$ and $a = 5$.

We already know two ways to do this. We could just **Author** and **Simplify** `3+3^2` and then do the same using `5+5^2`, or we could **Author** `F(x):=x+x^2` and then **Author** and **Simplify** `F(3)` and `F(5)`. A third method is to **Author** `a:=3`. This tells *DERIVE* to assign the value 3 to a. If we now **Author** and **Simplify** `a+a^2`, *DERIVE* will present the answer, 12, as in expression 8 of Figure 0.6. To evaluate at $a = 5$ we change the definition of a. **Author** `a:=5`, use the arrow keys to highlight $a + a^2$, and **Simplify** once more. The result appears in expression 10. If you want *DERIVE* to forget the definition of a and treat it as an arbitrary variable, **Author** `a:=`.

Editing expressions: Change $a + a^2$ to $a + 4a^2$.

Highlight $a + a^2$ and press [F3], which brings the highlighted expression to the author line where you may edit it. ([F4] will bring the highlighted expression to the author line enclosed in parentheses.)

In *DERIVE* versions 2.55 or later, the left and right arrow keys, [→] [←], perform dual functions: they move the cursor on the author line *or* they move the highlight on the screen. The [F6] key toggles between these two modes. In versions earlier than 2.55, the only way to move the cursor on the author line is with [Ctrl S] (left) and [Ctrl D] (right).

Use the left arrow key to move the cursor to the first occurrence of a. (If the arrow key does not make the cursor move, press [F6] and try again.) The [Insert] key toggles between typeover and insert modes. Make sure you are in the insert mode (Ins will appear in the lower right portion of the screen) and type `4`. The [Delete] and [Backspace] keys erase. When you are finished, press [Enter], and the new expression appears as expression 11.

Special Functions and Symbols

Keystrokes	Result
[Alt e] or `#e`	The number e (natural logarithm base)
[Alt p] or `pi`	The number π
[Alt i] or `#i`	The imaginary number i
`inf`	∞
`-inf`	$-\infty$
[Alt q] or `SQRT`	$\surd$

Practice Problems

1. Calculate the exact value of $\dfrac{2^8 + \sqrt{324}}{8!}$, and then obtain a decimal approximation.
   Answer: $\dfrac{137}{20160} \approx 0.00679563$

2. Factor $x^4 - 9x^2 - 4x + 12$. Answer: $(x-3)(x+2)^2(x-1)$

3. Solve $x^4 + 5x^3 + 14 = 3x^2 + 25x$. Answer: $x = -1 \pm 2\sqrt{2}$ and $x = \dfrac{-3 \pm \sqrt{17}}{2}$

4. Plot the graphs of x^2 and $\cos x$ on the same screen and estimate the coordinates of all points where the graphs cross. Answer: $(-0.824131, 0.679193)$, and $(0.824131, 0.679193)$

5. Determine if $2^{67} - 1$ is a prime number. (That is, does it have any whole number factors other than itself and 1.) Answer: It is not prime.

6. Let $f(x) = x^2 - 0.9$. Approximate the value of $f(f(f(f(1))))$. Answer: -0.888357

Quick Reference

Algebra Menu Commands

- **Author** allows you to input expressions.
- **Expand** expands the highlighted expression.
- **Factor** factors the highlighted expression.
- **Help** provides *DERIVE* User Manual references.
- **Jump** allows you to move to the expression number you choose.
- **soLve** solves the highlighted expression. If the expression is not an equation *DERIVE* attempts to find the zeros.
- **Manage Substitute** allows you to substitute values in the highlighted expression.
- **Options** allows you to change between **Exact** and **Approximate** modes and to set the number of digits *DERIVE* uses to make approximations.
- **Plot** opens a plot window or moves you to the plot window if one is already open.
- **Quit** ends a *DERIVE* session.
- **Remove** deletes the highlighted expression from the screen.
- **Simplify** simplifies the highlighted expression.
- **Transfer** allows you to save, load, or print files.
- **moVe** allows you to move the highlighted expression to a different place on the screen.
- **Window** allows you to open or close windows.
- **approX** gives the approximate value of the highlighted expression.
- The Tab key will move the cursor among fields.
- The escape key, Esc, will usually cancel a command, return you to previous menus, or get you out of an undesirable situation.

Plot Menu Commands

- **Algebra** returns to the algebra window.
- **Center** centers the picture at the graphing cursor.
- **Delete** gets rid of unwanted graphs.
- **Help** provides references to the *DERIVE* User Manual.
- **Options** lets you change graphics options.
- **Plot** starts plotting. ([Esc] stops it.)
- **Quit** ends a *DERIVE* session.
- **Scale** allows you to manually set the scale.
- **Window** allows you to close windows or open new ones.
- **Zoom** allows you to zoom in and out. Alternatively, [F9] zooms in and [F10] zooms out.
- **Arrow keys** move the graphing cursor. [Ctrl →], [Ctrl ←], [PageUp], and [PageDown] move the graphing cursor in larger jumps.

Editing

- **Arrow keys** move the highlight among expressions on the algebra screen and also move the cursor on the author line. The [F6] key toggles between the two modes. In a plot window the arrow keys move the graphing cursor.
- [F3] brings the highlighted expression to the author line.
- [F4] brings the highlighted expression to the author line enclosed in parentheses.
- [Insert] toggles between insert and typeover modes.
- An expression can be referred to by # followed by its number.

Chapter 1

Functions and Graphs

<u>New *DERIVE* topics</u> •Plotting graphs •Solving equations exactly •Solving equations approximately •Plotting lists of functions •Plotting equations
<u>Calculus concepts</u> •Function •Graph •Zero •Domain •Range •Trigonometric functions

Solved Problem 1.1: Domains, ranges, and zeros of functions

<u>New *DERIVE* Lessons</u> •Solving equations exactly •Toggling between exact and approximate modes •Solving equations approximately •Graphing functions •Moving the graphing cursor •Zooming in and out

For each of the following functions, plot the graph and find the zeros, domain, and range.

(a) $f(x) = \dfrac{x^3 - 5x^2 - x + 5}{10}$

(b) $f(x) = x^4 - \sqrt{1-x}$

<u>Solution to (a)</u>: First **Author** `(x^3-5x^2-x+5)/10` which appears in Figure 1.1 as expression 1. Press [P] to **Plot** and [Enter] twice to acknowledge **Beside** and **40**. Press [P] again, and *DERIVE* presents the graph in window 2 of Figure 1.1. (In this figure we have split the plot window to accommodate an extra graph. It is not necessary for you to do this, but if you wish to learn more about window splitting, see your *DERIVE* manual or Section 3 of Appendix I.) A cubic polynomial may have as many as three roots, but the graph shows only two. We zoom out with [F10] twice to see the graph in window 3. (The resulting scale in window 3 is x:5 y:5.) With this scale the graph can be seen to cross the x-axis once more indicating a third zero; the first view just didn't show enough of the graph. This picture also suggests that the domain and range of f are both $(-\infty, \infty)$.

There is an important lesson here that applies to any graphics software: The initial view may not display all the interesting information about the graph of a function, and it will often be necessary for you to make intelligent adjustments to the picture to obtain the information you are after.

To find the zeros, return to your calculations by pressing [A] for **Algebra**. Press [L] for **soLve**, and we see the zeros in expressions 2, 3, and 4 of Figure 1.1.

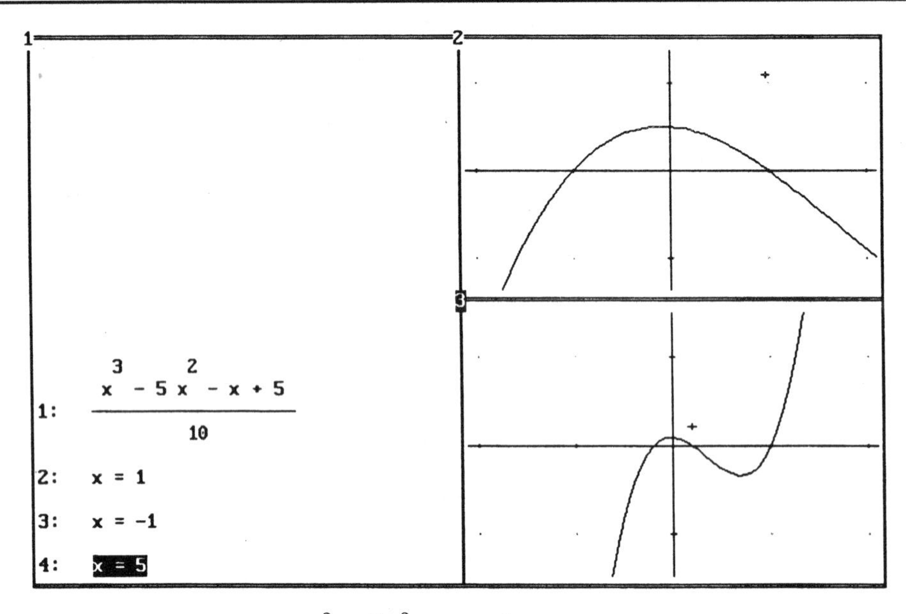

Figure 1.1: **The graph of** $\dfrac{x^3 - 5x^2 - x + 5}{10}$. **Scale in window 3, x:5 y:5**

Solution to (b): **Author** `x^4-SQRT(1-x)`. (You may use either [Alt q] or type `SQRT` to make a square root symbol.) We **Plot Beside** as in (a), and the graph appears as in window 3 of Figure 1.2. We do not need a computer to find the domain of f; $1 - x$ must be nonnegative to avoid taking the square root of a negative number. Thus the domain consists of all real numbers less than or equal to 1, and the graph confirms this. It is not so easy to see what the range is. The graph in Figure 1.2 indicates that function values may be arbitrarily large but can be no smaller than the "bottom" of the graph; we use the arrow keys to move the graphing cross to this spot. Once the cross is properly positioned, we can read the coordinates of its location from the lower left corner of the screen; Cross: x:−0.5 y:−1.1562 (See Figure 1.2). Thus the range of f is approximately $[-1.1562, \infty)$.

To find the zeros, we return to our calculations using **Algebra**. When we ask *DERIVE* to **soLve** as we did in (a), we see in expression 2 of Figure 1.2 that it just returns the input with an "= 0" tacked onto the end. This indicates that *DERIVE* does not know how to find the zeros of this function exactly.

In the plotting tutorial in the Introduction, we learned how to approximate the zeros of a function by zooming in on points where the graph crosses the x-axis. We will show here a method that provides greater accuracy. The first step is to set *DERIVE* to its approximate mode using **Options Precision Approximate** [Enter]. (Press [O] for **Options** and you will be presented with a new menu, one of whose items is **Precision**. Press [P], and you will be prompted with

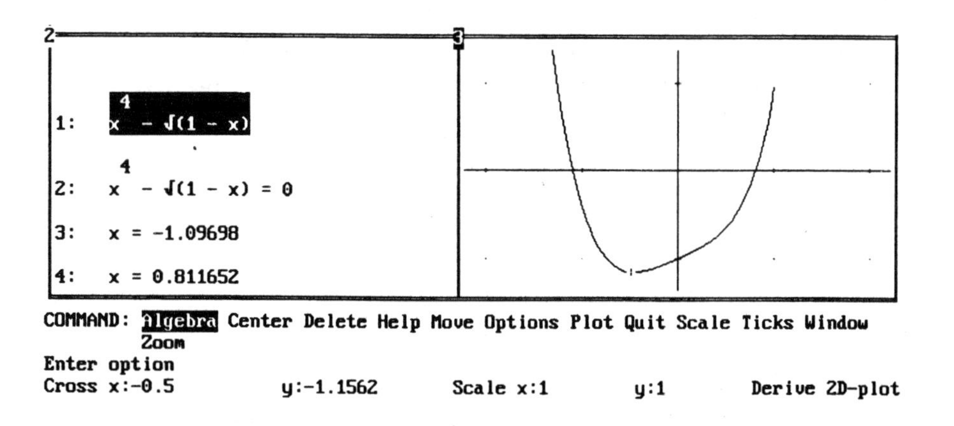

Figure 1.2: **Graph of $x^4 - (1-x)^{\frac{1}{2}}$**

OPTIONS PRECISION: Mode: Approximate Exact Mixed Digits:6. Press A for **Approximate** and Enter.) In this mode, *DERIVE* gives approximate rather than exact answers.

Now if we **soLve**, we are prompted with SOLVE: Lower:-10 Upper:10. *DERIVE* is offering to look for a zero on the range from -10 to 10, *and it will find at most one zero on this interval no matter how many there may actually be.* Thus, we should adjust this range to make certain it contains exactly one zero. The graph shows that there is one zero in the interval $[-2,-1]$ and another in $[0,1]$. We change the range to Lower:-2 Upper:-1 and Enter. (The Tab key will move to the Upper field.) *DERIVE* presents us with the solution in expression 3 of Figure 1.2. To find the second zero, use the arrow keys to highlight expression 2 once more and **soLve** setting the range to Lower:0 Upper:1. The answer is in expression 4. Before proceeding, return to the exact mode using **Options Precision Exact** Enter.

You may be interested to know that when *DERIVE* solves equations in the approximate mode, it uses a technique known as the *bisection method*, which may be discussed in your calculus text.

A note about inequalities: The information we have found in (b) is actually enough to solve the seemingly harder problem, $x^4 < \sqrt{1-x}$. This inequality is true when the graph of $x^4 - \sqrt{1-x}$ is below the x-axis. From the graph in Figure 1.2 we see that this happens between the two points where the graph crosses the axis. Thus, the solution is $-1.09698 < x < 0.811652$.

A note about Mixed Mode: For equations that *DERIVE* cannot solve exactly, there is an alternative method to the one presented above that sometimes works. You can set *DERIVE* to Mixed Mode using **Options Precision Mixed** Enter. If you **soLve** in this mode, *DERIVE* will give those approximate solutions that it is able to find without further help. For quadratic, cubic, and most quartic equations, *DERIVE* will approximate all of the zeros, but for other equations, it does not work so well. For example, if you try to find the zeros for (b) of Solved Problem 1.1, *DERIVE* will find the positive root but not the negative one. In general, if you use Mixed Mode to solve equations you are advised to check the graph to be sure you have all the solutions.

Solved Problem 1.2: A floating ball

According to *Archimedes' law*, the weight of water that is displaced by a floating object is equal to the weight of the object. If a ball of radius r floats in water, then the volume of the submerged portion of the ball is given by $\pi x^2 \left(r - \frac{x}{3}\right)$ where x is the submerged depth.
A ball of radius 2 feet weighs 76 pounds. If it is placed in pure water, which has a density of 62.4 pounds per cubic foot, how much of the ball's diameter is beneath the surface of the water?

Solution: Let x denote the submerged depth of the ball. Then the volume of water displaced is the volume of the portion of the ball beneath the surface.

$$\text{volume of water displaced} = \pi x^2 \left(2 - \frac{x}{3}\right) \text{ cubic feet}$$

Water weighs 62.4 pounds per cubic foot, so that the *weight* of water displaced is $62.4\pi x^2 \left(2 - \frac{x}{3}\right)$ pounds. Equating this to the weight of the ball, we obtain $62.4\pi x^2 \left(2 - \frac{x}{3}\right) = 76$.

We need to solve this equation for x. **Author** `62.4 pi x^2(2-x/3)=76` as seen in expression 1 of Figure 1.3. If we **soLve**, *DERIVE* presents us with the complicated expressions 2, 3, and 4. Highlight each one in turn and **approX** to get the results in expressions 5, 6, and 7. We are looking for one solution, but *DERIVE* gave us three. Notice that expression 5 is negative and expression 6 is larger than the diameter of the ball. Thus the only reasonable answer is $x = 0.458111$ feet. The other two are indeed solutions to the cubic equation in expression 1, but that equation only describes the physical situation on the range $0 \le x \le 4$. An alternative method for solving this problem is to change to the approximate mode and **soLve** on the range `lower:0 Upper:4`.

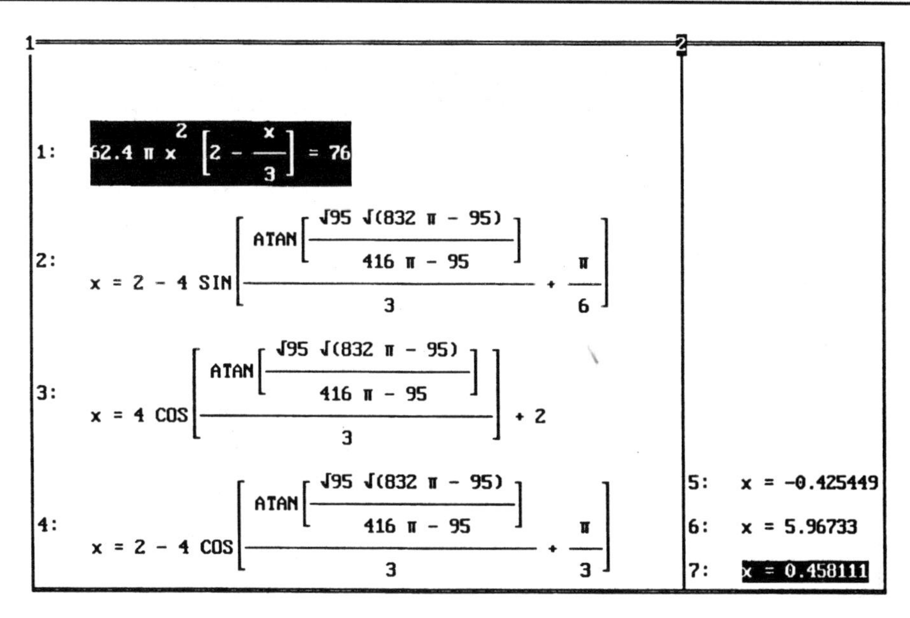

Figure 1.3: **A floating ball**

Trigonometric Functions: A Note about Radians versus Degrees

DERIVE's syntax for the six trigonometric functions is standard, but radian measure is assumed. Thus *DERIVE* reads `SIN(30)` as the sine of 30 radians. If you want the sine of 30 degrees, you must use `SIN(30 deg)`.

Solved Problem 1.3: Periodicity of trigonometric functions

Plot the graphs of $\cos^2 x$ and $\cos(x^2)$. Determine if they are periodic, and if they are, find the periods.

Solution: **Author** `COS^2 x` as seen in window 1 of Figure 1.4. *DERIVE* chooses to write this function as $\cos(x)^2$ rather than $\cos^2(x)$ as we might have expected. Both mean $(\cos x)^2$. Now **Plot Under** so we can view the graph on a large domain. The graph in window 4 of Figure 1.4 appears to repeat every π units. To verify this, **Author** `COS ^2 (x+pi)` and **Plot**. You will see that the new graph overlays the old one exactly, and we conclude that the period is indeed π. You should be able to use elementary facts about the cosine function to verify this algebraically.

Now **Author** and **Plot** `COS(x^2)`. The graph appears in window 5 of Figure 1.4. It does not look periodic, and it isn't. (Explain why.)

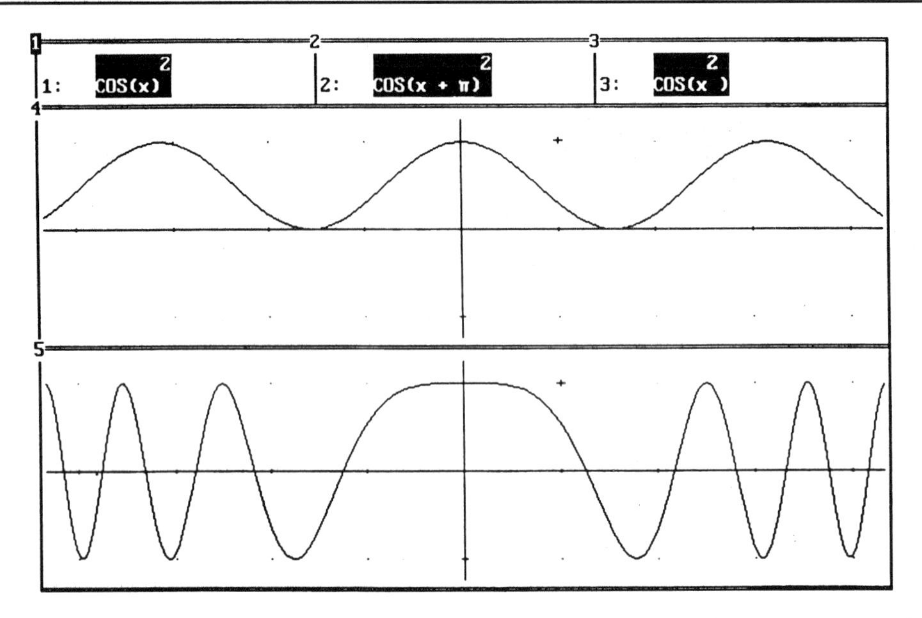

Figure 1.4: **Graphs of** $\cos^2 x$ **and** $\cos(x^2)$

Practice Problems

1. Solve the equation $x^3 + \sqrt{x-1} = 5$. Answer: $x = 1.61541$

2. Solve the inequality $|x^2 - 3| < 6x$. Note: The absolute value bar is usually found on the same key with the back-slash "\." Answer: $0.464101 < x < 6.46410$

3. Find the range of $x^2 - x - 2$. Answer: $[-2.25, \infty)$

4. Plot the graphs of $\tan x$ and $\sec x$.

Laboratory Exercise 1.1

Zeros, Domain, and Range

Name ______________________________ Due Date ____________

Plot the graphs of the following functions and find the zeros, domain, and range. Explain how you got your answers.

1. $f(x) = 1 + 8x^2 - x^4$

2. $g(x) = \sqrt{3x - 1} + \sqrt{6 - x} - x$

3. $h(x) = x^5 - x^3 - 800x + 700$

4. $k(x) = \dfrac{x}{x^3 + x + 1}$

Laboratory Exercise 1.2

A Ball Floating on the Ocean

Name ______________________ Due Date __________

The ball from Solved Problem 1.2 is moved to the ocean. Sea water has a density of 63.9 pounds per square foot.

1. How much of the diameter of the ball will be beneath the surface of the ocean?

2. How much should the ball weigh so that it floats completely submerged? (Its top will be at the surface of the water.)

3. Consider a ball of weight w. Plot the graph of the submerged depth versus the weight of the ball. Is it a straight line?

Laboratory Exercise 1.3

Bald Eagles

Name ______________________________ Due Date __________

A breeding group of bald eagles is released into a protected area in New Mexico and is carefully monitored. The population t years after release is given by

$$P(t) = \frac{40}{1 + 3(0.67)^t}$$

1. How many eagles were released?

2. It is thought that the eagle population can maintain itself without outside interference once the population reaches 30 individuals. How long must a care program for the eagles be maintained?

3. Plot the graph of P versus t and explain in words how the population varies with time.

4. Estimate the time when the population is growing at the fastest rate. Explain how you got your answer.

5. How many eagles can the environment in the protected area support? (Hint: Look at the graph of P for large values of t.)

Chapter 2

Limits

New *DERIVE* topics •Defining functions •Making lists •Calculating limits •Setting precision •Setting the graphing scale •Plotting piecewise defined functions •Plotting vertical lines •Cleaning up fuzzy graphs
Calculus concepts •Limits •Limits at infinity •One-sided limits •Horizontal and vertical asymptotes •Definition of limit

Solved Problem 2.1: Graphical and numerical estimation of limits

New *DERIVE* Lessons •Defining functions •Making lists •Setting the graphing scale

Plot the graph and compute function values to estimate the following limits. Check your answers with *DERIVE*.

(a) $\lim_{x \to 0} \dfrac{\sin(3x)}{7x}$

(b) $\lim_{x \to 1} \sin\left(\dfrac{1}{1-x}\right)$

(c) $\lim_{x \to \infty} (\sqrt{x^2 + x + 1} - x)$

Solution to (a): **Author** `SIN(3x)/(7x)` and **Plot Beside** to get the graph in window 2 of Figure 2.1. We are interested in what happens to the function near $x = 0$; that is where the graph appears to cross the y-axis. (Note: The function is not defined at $x = 0$, and there is actually a hole in the graph at that point.) Use the arrow keys to move the graphing cross close to that point, **Center**, and use F9 to zoom in twice. (You may need to adjust the graphing cursor and **Center** again.) The resulting graph appears in window 3 of Figure 2.1, and we see the location of the graphing cross at the bottom of the screen, Cross: x:0 y:0.4281. We conclude that the limit is approximately 0.4281. We could achieve greater accuracy by zooming in further.

To make a numerical estimate of the limit, we define $f(x)$ to be $\dfrac{\sin(3x)}{7x}$ and look at function values near $x = 0$. **Author** `F(x):=#1`. (Note: Be sure to use "`:=`" which denotes a definition and not just "=" which denotes an equation.) Now we will examine $f(-0.002)$, $f(-0.001)$, $f(0.001)$, and $f(0.002)$. *DERIVE* can calculate several things at once if they are put in a list enclosed by

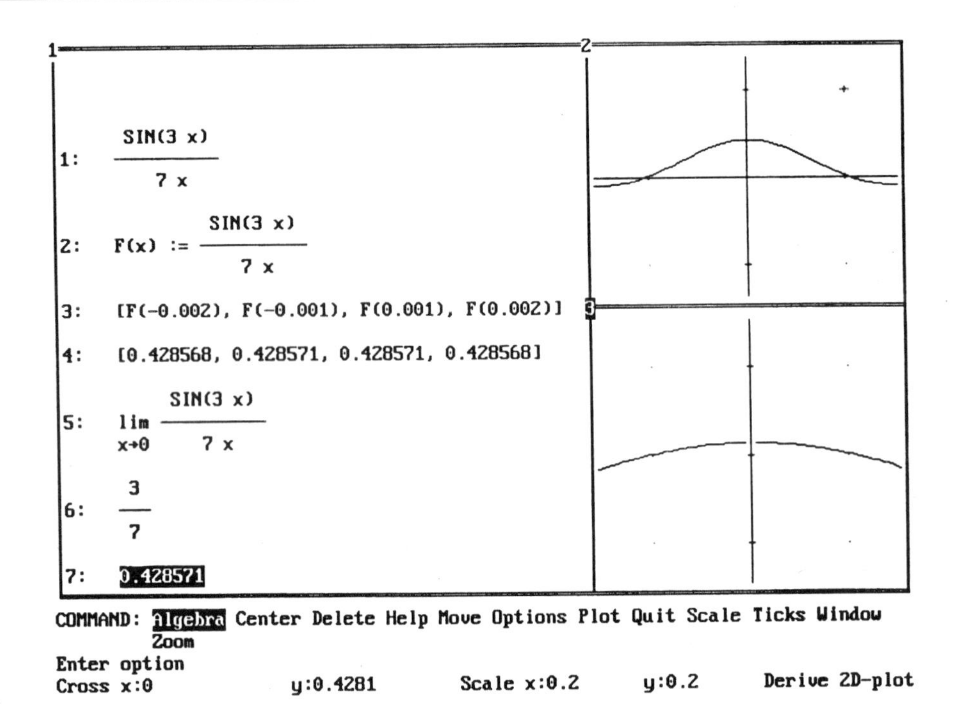

Figure 2.1: $\lim_{x \to 0} \dfrac{\sin(3x)}{7x}$

square brackets and separated by commas. Thus we **Author** `[F(-0.002),F(-0.001),F(0.001),F(0.002)]`.

To obtain the function values in expression 4 of Figure 2.1, use **approX**. From this list, it appears that the limit is approximately 0.428571, and this is close to our graphical estimate. You can improve accuracy by looking at values of x even closer to zero, but care must be taken. See **A Note on Accuracy** following this Solved Problem.

To calculate the exact value of this limit, highlight expression 1 and use **Calculus Limit** setting the CALCULUS LIMIT: Point: field to `0`. (Press C for **Calculus**, and we are presented with a new menu, one of whose items is **Limit**. Press L and then Enter at each of the prompts expression: # and variable: x. Finally, you will see the prompt Point: 0. This is correct, so press Enter once more.) *DERIVE* presents us with the notation for the limit in expression 5 of Figure 2.1. **Simplify** to get the answer, $\frac{3}{7}$, in expression 6. We can **approX** this to get the decimal approximation in expression 7 for comparison with our graphical and numerical estimates above.

Solution to (b): **Author** `SIN(1/(1-x))` and **Plot Beside**. This time we are interested in what happens near $x = 1$. In window 2 of Figure 2.2 we see that the graph appears to oscillate wildly near $x = 1$ and does not approach any single value. In window 3 we have **Center**ed at the point $(1, 0)$ and zoomed in twice with F9. The erratic behavior is even more apparent, and we conjecture that the limit does not exist.

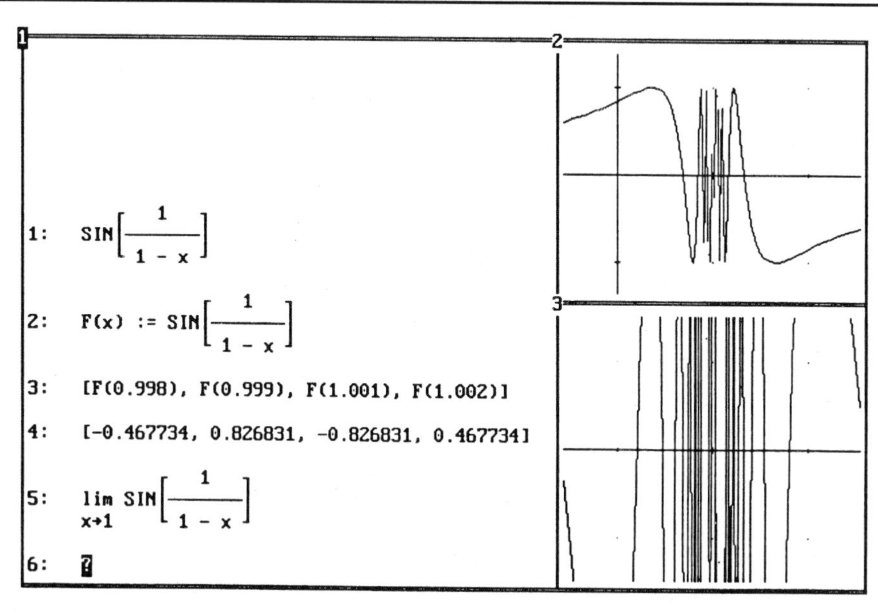

Figure 2.2: $\lim_{x\to 1} \sin\left(\frac{1}{1-x}\right)$

To corroborate this numerically, we proceed as in Part (a). **Author** `F(x):=#1` and then **Author** `[F(0.998), F(0.999), F(1.001), F(1.002)]`. When we **approX**, *DERIVE* presents us with the list in expression 4. The numbers don't seem to be close to a single value, and we are led once again to suspect that the limit does not exist.

To get *DERIVE* to calculate the limit, highlight expression 1 and use **Calculus Limit**. This time set the Point: field to `1`. When we **Simplify**, *DERIVE* returns a question mark as seen in expression 6. (This does not prove that the limit does not exist; it simply means that *DERIVE* does not know how to calculate it.) The graphical and numerical evidence has in fact led us to the right conclusion; the limit does not exist.

Solution to (c): **Author** `SQRT(x^2+x+1)-x` and **Plot Beside**. The graph appears in window 2 of Figure 2.3. We are interested in what happens for large values of x, so in window 3 we have changed the scale. From the plot window, we press S, for **Scale** and we are prompted with

Scale x:1 y:1. We change this to Scale x:100 y:1 and press Enter. (We left the y scale at its default value of 1. If you want to change it, the Tab key will move you to that field.) We have used the arrow keys to move the graphing cross toward the right-hand end of the graph, and we read the location of the graphing cross from the bottom of Figure 2.3, Cross x:273.611 y:0.5. Thus it appears that the limit is near 0.5.

As in (a) and (b), we **Author** `F(x):=#1`. We are interested in what happens to the function when x is large, so we **Author** and **approX** `[F(100), F(1000), F(10000), F(100000)]`. The result in expression 4 of Figure 2.3 verifies our estimate of 0.5.

Finally, we highlight expression 1 and use **Calculus Limit** setting the Point: field to `INF` to obtain expression 5. When we **Simplify**, *DERIVE* reports that the limit is $\frac{1}{2}$.

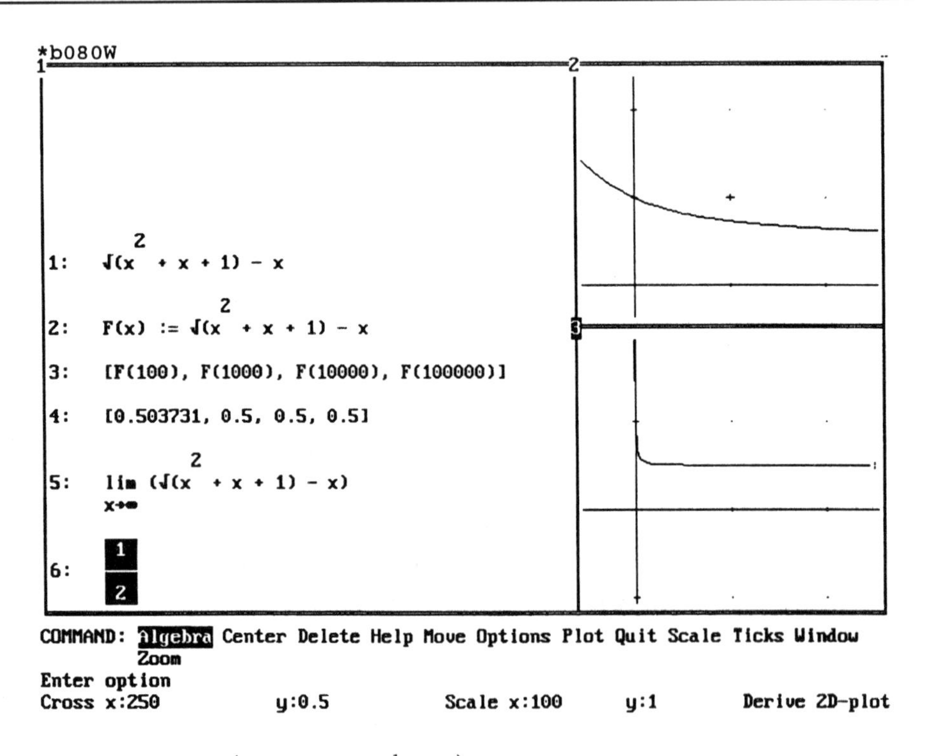

Figure 2.3: $\lim_{x\to\infty}\left((x^2+x+1)^{\frac{1}{2}}-x\right)$. **Scale in window 3, x:100 y:1**

A Note on Accuracy

New *DERIVE* Lessons •Setting precision

We suggested in Solved Problem 2.1 that the precision of numerical approximations of limits can be improved by looking at function values very near the limit value. This is true, but care must be taken not to exceed the number of digits being used to make approximations.

We illustrate the point with $\lim_{x\to 0}(1+x)^{\frac{1}{x}}$. The value of this limit is a special number, $e \approx 2.71828182845$, that will be encountered as the course progresses. **Author** `F(x):=(1+x)^(1/x)` as seen in expression 1 of Figure 2.4. We have used the VECTOR command as described in Chapter 8.2 of the *DERIVE* User Manual to make a table of values of $f(x)$ for $x = 0.1, 0.01, 0.001, 0.0001, 0.00001$, and 0.000001 in expression 3. The first four rows of the table show that $f(x)$ is getting close to the appropriate limit, but the last two seem to give evidence to the contrary.

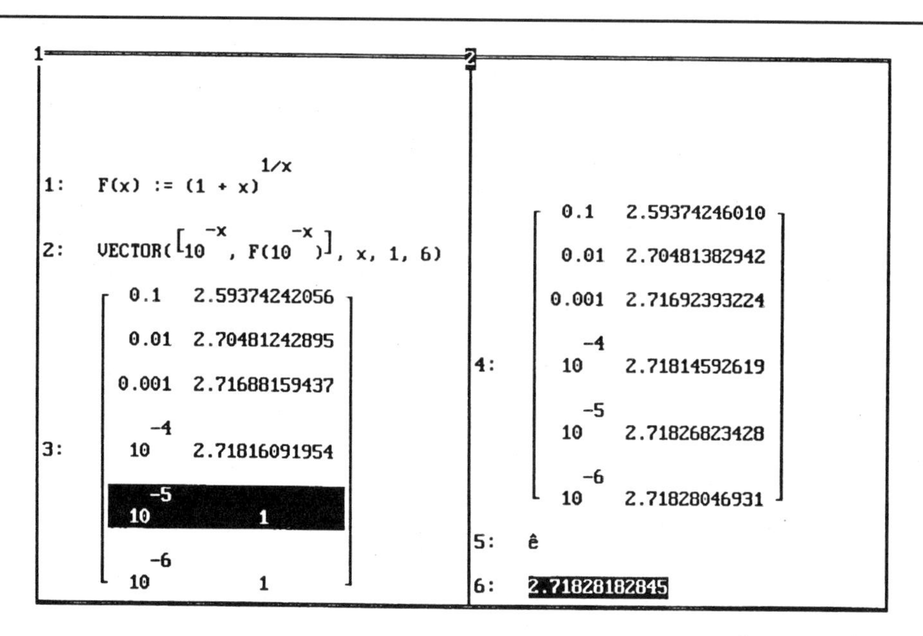

Figure 2.4: **Tables of values for** $\lim_{x\to 0}(1+x)^{\frac{1}{x}}$

In its default mode, *DERIVE* uses six digits to make approximations, but round-off error causes this to be insufficient for a good approximation to $f(10^{-5})$. We can fix that by changing the precision using **Options Precision** [Tab] and then setting Digits: to whatever number we choose. To make the table in expression 4 of Figure 2.4 we set Digits: 12, highlighted expression 2 once more, and used **approX**. The last two rows of the new table are consistent with the first four.

A Warning about Unusual Answers

DERIVE may occasionally present you with answers involving functions or constants that you are not familiar with. For example, if you ask *DERIVE* to calculate $\lim_{x\to 0}\frac{(1+x)^{\frac{2}{x}}(2^x-1)}{x}$, it will present the answer as $e^2 \ln 2$. This is the correct answer, but depending on the calculus text you are using, neither the number e nor the function LN may have been introduced. When this occurs, you may **approX** and report the decimal approximation, in this case, 5.12170.

Solved Problem 2.2: One-sided limits

New *DERIVE* Lessons •Calculating one-sided limits •Plotting piecewise defined functions

For each of the following functions, make a graph showing the behavior of the function near the given point. Use *DERIVE* to calculate the left-hand and right-hand limits and determine if the (two-sided) limit exists.

(a): $f(x) = \dfrac{x^3-1}{|x^2-1|}$ Limit point 1

(b): $g(x) = \begin{cases} \frac{\sin x}{2x} & \text{if } x < 0 \\ \frac{1}{x+2} & \text{if } x > 0 \end{cases}$ Limit point 0

Solution to (a): First, we **Author** `(x^3-1)/|x^2-1|`. (On a standard keyboard, the absolute value bar is on the same key with the back-slash "\".) When we **Plot Beside** we see that the graph in window 3 of Figure 2.5 breaks at $x = 1$, and we suspect that the left and right-hand limits at $x = 1$ are not the same.

Use **Algebra** to return to the calculations. To get the left-hand limit, we use **Calculus Limit**, but at the prompt

CALCULUS LIMIT: Point: 0 (Both) Left Right

set Point: `1`, and then use the Tab key to move to the right-hand side of the prompt. The parentheses around "Both" indicates that *DERIVE* is ready to calculate a two-sided limit. Press L for **Left**, and this part of the prompt will change to Both (Left) Right. Now press Enter and *DERIVE* will display the notation for the left-hand limit in expression 2 of Figure 2.5. **Simplify** to get the answer in expression 3.

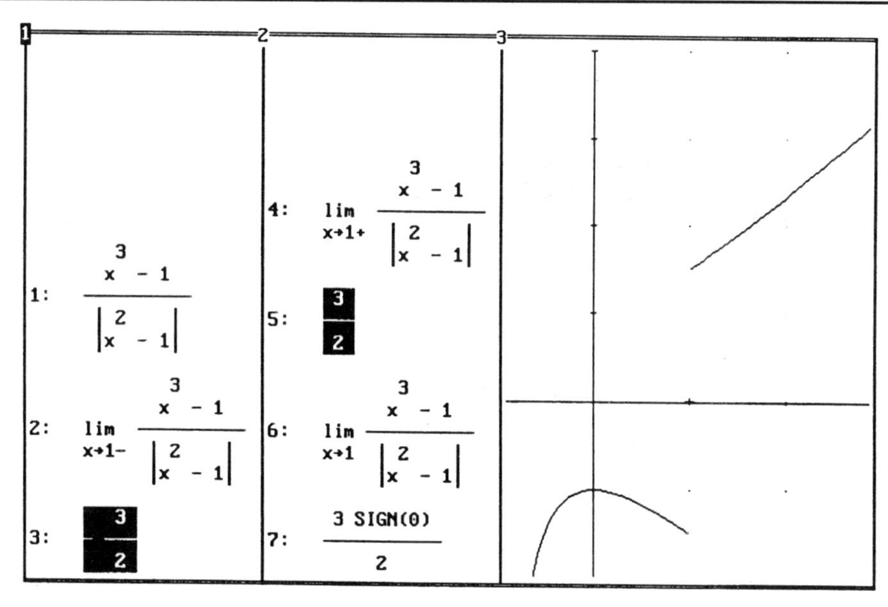

Figure 2.5: $\lim_{x\to 1^-} \dfrac{x^3-1}{|x^2-1|}$ **and** $\lim_{x\to 1^+} \dfrac{x^3-1}{|x^2-1|}$

To get the right-hand limit, highlight expression 1 again and use **Calculus Limit** as before. This time, be sure the (Right) option is selected. **Simplify** to get the right-hand limit in expression 5.

Since the left-hand and right-hand limits do not agree, we know that the two-sided limit does not exist, but let's see what *DERIVE* says about it. Proceed as above being sure the parentheses enclose "Both." When we **Simplify**, *DERIVE* presents us with expression 7 which involves the SIGN function. This function is defined to be 1 for $x > 0$ and -1 for $x < 0$; it is not defined at $x = 0$. Thus, *DERIVE's* answer is consistent with our conclusion that the limit does not exist.

Solution to (b): There is a trick that allows us to plot functions with different definitions on different intervals such as the one we are dealing with here; it uses *DERIVE's* CHI function.

$$\text{CHI}(a, x, b) = \begin{cases} 0 \text{ if } x < a \\ 1 \text{ if } a < x < b \\ 0 \text{ if } x > b \end{cases}$$

The graph of CHI$(0, x, 1)$ appears in window 2 of Figure 2.6. We will make things simple so that you need not be overly concerned about how this function works. **Author** `ON(a,b):=CHI(a,x,b)`. Now, we want our function to be $\dfrac{\sin x}{2x}$ on the interval $(-\infty, 0)$ and $\dfrac{1}{x+2}$ on the interval $(0, \infty)$. Thus we **Author** `SINx/(2x) ON(-INF, 0) + 1/(x+2) ON(0,INF)`.

Plot Beside to get the graph in window 3 of Figure 2.6. It appears from the graph that the

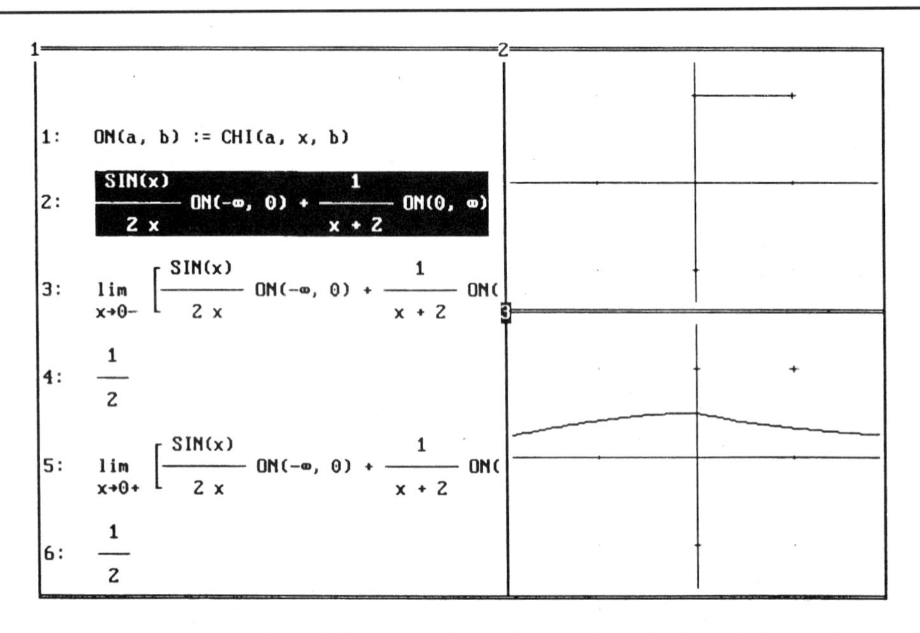

Figure 2.6: **One-sided limits of a piecewise defined function**

left-hand and right-hand limits at zero are the same. This is verified in expressions 4 and 6, and we conclude that the two-sided limit exists and is the common value, $\frac{1}{2}$. If you ask *DERIVE* to calculate the two-sided limit, it will report the correct answer.

Solved Problem 2.3: Asymptotes

New *DERIVE* Lessons •Plotting vertical lines

Plot the graph of $\frac{x^3+x+2}{5x^3-9x^2-4x+4}$. Find the horizontal and vertical asymptotes and include the asymptotes in the picture.

Solution: **Author** `(x^3+x+2)/(5x^3-9x^2-4x+4)` and **Plot Beside**. The graph is in window 2 of Figure 2.7.

To find the horizontal asymptotes, we calculate the limits as x goes to ∞ and $-\infty$. From expressions 3 and 5 we see that the result in each case is $\frac{1}{5}$. With expression 5 highlighted, we **Plot** once more to include the horizontal asymptote in the picture.

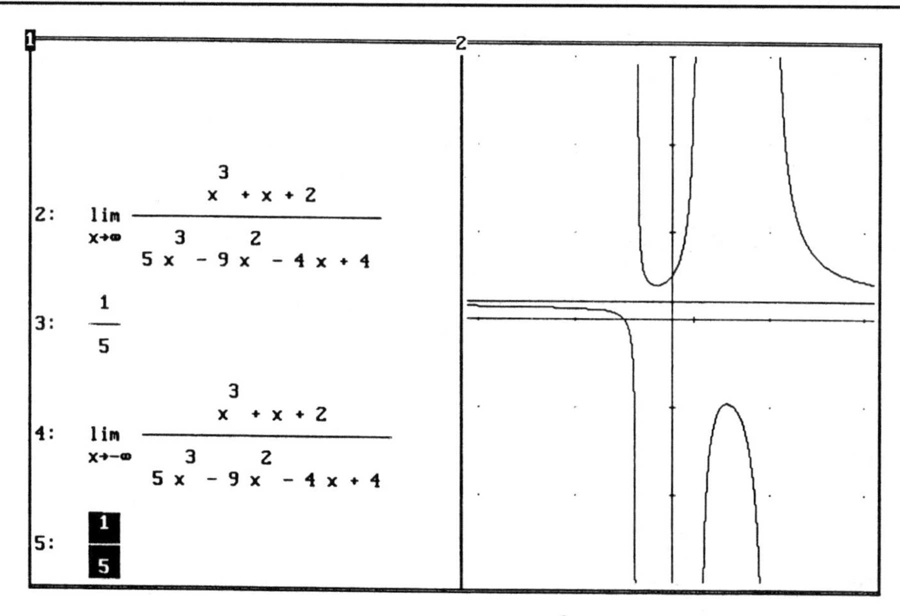

Figure 2.7: **Horizontal asymptotes for** $\dfrac{x^3+x+2}{5x^3-9x^2-4x+4}$. **Scale x:2 y:1**

To find the vertical asymptotes, we need to find the zeros of the denominator. Use the arrow keys to highlight the denominator of expression 1 (not shown). Now **Author** [F3] and $5x^3 - 9x^2 - 4x + 4$ appears in expression 6 of Figure 2.8.

We **soLve** to obtain the solutions in expressions 7, 8, and 9. These are the vertical asymptotes we want to plot, but a vertical line is not the graph of a function. To plot these lines, we use a trick with *DERIVE's parametric plot* feature. **Author** `[2,t]` and **Plot** as usual. We see the prompt

Plot: Min: -3.14159 Max: 3.14159

DERIVE is offering to plot the ordered pairs $(2, t)$ as t ranges from -3.14159 to 3.14159. Our graph extends vertically from about -3 to 3, and so we simply accept the default values by pressing [Enter]. (If you want to change these numbers, you will have to use the [Tab] key to get to the Max: field.) Repeat this procedure for the solutions in expressions 8 and 9 of Figure 2.8. The completed picture is in window 2.

Suggestion: You can use the [F3] key to save some typing when you make expressions 11 and 12. Use the arrow keys to highlight the right-hand side of the equation in expression 8 as seen in Figure 2.8. Now **Author** `[`[F3]`,t]` [Enter], and you will see expression 11.

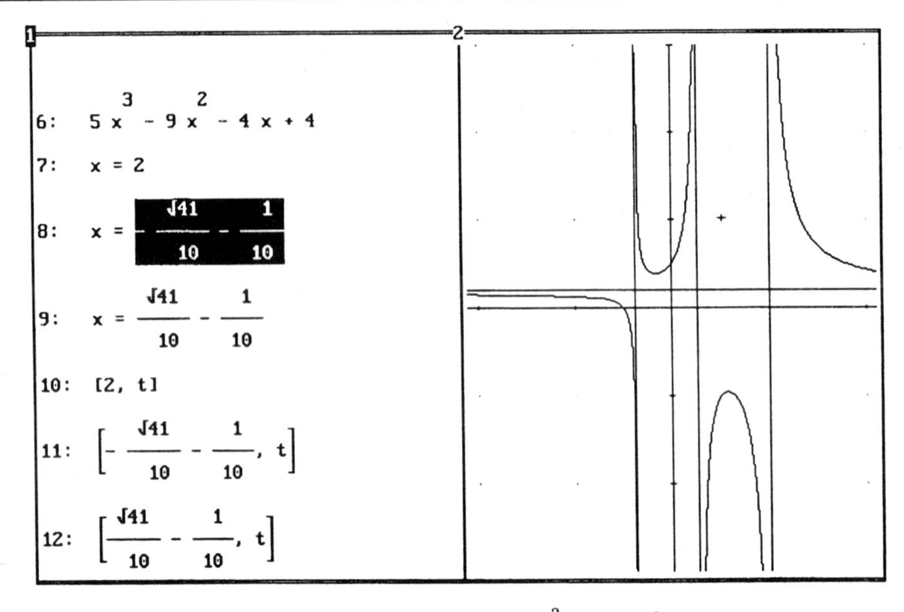

Figure 2.8: **Vertical asymptotes for** $\dfrac{x^3+x+2}{5x^3-9x^2-4x+4}$. **Scale x:2 y:1**

Solved Problem 2.4: Illustrating the definition of the limit

New *DERIVE* Lessons •Cleaning up fuzzy graphs

Use *DERIVE* to verify that $\lim_{x\to\pi} \dfrac{\sin(3x)}{\pi - x} = 3$. Find a value of δ so that if $0 < |x-\pi| < \delta$, then $\left|\dfrac{\sin(3x)}{\pi - x} - 3\right| < 0.001$.

Solution: First **Author** `SIN(3x)/(pi - x)` and then use **Calculus Limit** setting the Point: field to `pi`. **Simplify**, and the limit is verified in expression 3 of Figure 2.9.

To find δ, we need to solve the inequality $\left|\dfrac{\sin(3x)}{\pi - x} - 3\right| - 0.001 < 0$. **Author** `|#1-3|-0.001` and **Plot Beside**. Move the graphing cross near $(\pi, 0)$ and **Center** to see the graph in window 2 of Figure 2.9. With the default scale, it is not clear that the graph ever dips below the x-axis, but that is not surprising since the smallest it can possibly be is -0.001. (Why?) In window 2 we have centered at the origin and used F9 to zoom in six times, **Center**ing and adjusting the graphing cross as necessary. This shows the graph dipping below the x-axis, but the fuzzy nature of the graph indicates that we have made the scale too small to get an accurate plot with the default precision of six digits. From the plot window we use **Options Precision** Tab and set Digits: 12. This produces the smoother graph in window 4. With this view we see that the graph dips below the x-axis just to the left of $x = \pi$ and *stays there* until it crosses again just to the right of $x = \pi$. We need to locate these crossing points. We use **Algebra** to return to our calculations and toggle to the approximate mode using **Options Precision Approximate** Enter. Now with expression 4 highlighted, we **soLve** setting Lower: 3 Upper: pi. The result appears in expression 5. We highlight expression 4 once more and **soLve** on the range Lower: pi Upper: 4. In expressions 8 and 10 we have calculated the distance from the crossing points to π. Any positive value of δ less than both of these will work. For example, we can take $\delta = 0.01$.

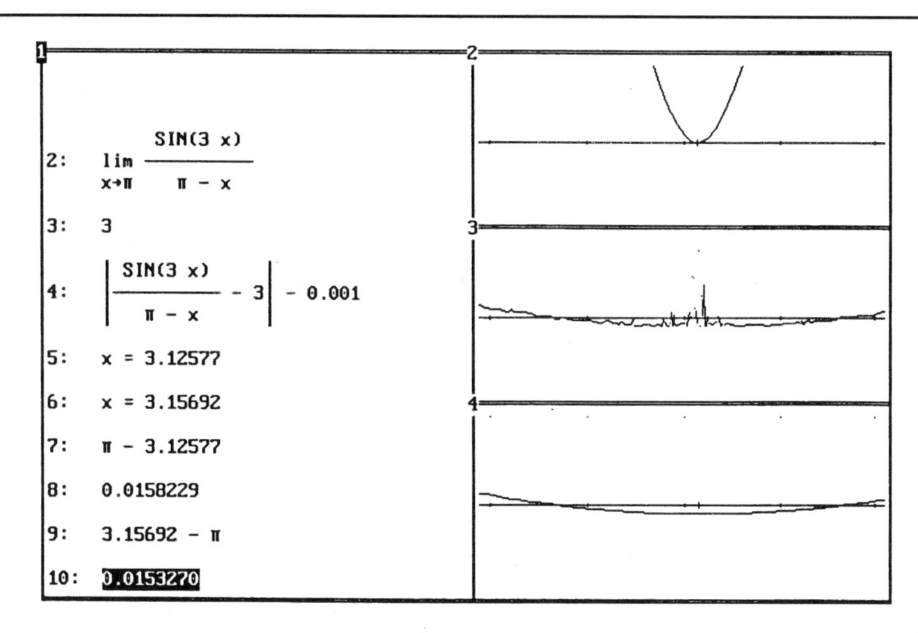

Figure 2.9: **Illustrating the definition of the limit**

Practice Problems

1. Use graphs to estimate $\lim\limits_{x \to 0} \dfrac{3^x - 1}{x}$. Answer: 1.09861

2. Use *DERIVE* to calculate the exact value of $\lim\limits_{x \to \pi} \dfrac{\sin(3x) - \sin(5x)}{\sin(2x) + \sin(4x)}$. Answer: $-\dfrac{1}{3}$

3. Use *DERIVE* to calculate the exact value of $\lim\limits_{x \to \infty} \dfrac{\cos x}{x}$. Answer: 0

4. Calculate $\lim\limits_{x \to 0+} \left(1 + \dfrac{1}{x}\right)^x$. Answer: 1

5. Plot the graph of $f(x) = \begin{cases} \sin x \ , \text{ if } \ x < 1 \\ \\ \cos x \ , \text{ if } \ x > 1 \end{cases}$

6. Solve the inequality $\left|\dfrac{2x^4 - 2x^3 + x - 1}{x^3 + 7x - 8} - \dfrac{3}{10}\right| < 0.001$. Answer: $0.998035 < x < 1.00195$

7. Solve the inequality $\left|\dfrac{3x^2 + 1}{x^2 + 2} - 3\right| < 0.001$. Answer: $x > 70.695$ or $x < -70.695$

8. Find the vertical asymptotes of $\dfrac{x}{x^3 + x - 1}$. Answer: $x = 0.682328$ is the only one.

9. Plot the graph of the vertical line $x = 3$ using *DERIVE*.

Laboratory Exercise 2.1

Graphical and Numerical Estimation of Limits

Name ______________________________ **Due Date** ____________

Use graphical and numerical techniques to estimate the value of each of the following limits and explain how you got your estimates. Check your answer where possible by using *DERIVE* to calculate the exact value of the limit.

1. $\lim_{x \to 0} (1 + 2x)^{\frac{3}{x}}$

2. $\lim_{x \to 1} \dfrac{2^x - 2}{x^2 - 1}$

3. $\lim_{x \to 0} x \sin\left(\dfrac{1}{x}\right)$

4. $\lim_{x \to \infty} ((x^3 + x^2 + 1)^{\frac{1}{3}} - x)$

5. $\lim_{x \to \infty} \left(1 + \frac{2}{x}\right)^x$

6. $\lim_{x \to \infty} \sin(\sin x)$

Laboratory Exercise 2.2

When Graphical Estimates Lead You Astray

Name ______________________________ Due Date ____________

Let $f(x) = x^{-10} 1.0001^{-\frac{1}{\sqrt{|x|}}}$.

1. Estimate $\lim_{x \to 0} f(x)$ graphically.

2. Estimate $\lim_{x \to 0} f(x)$ numerically.

3. Ask *DERIVE* to calculate $\lim_{x \to 0} f(x)$.

4. How do you reconcile Part 3 with your estimates in Parts 1 and 2?

5. Make a hand-drawn graph of f that shows its behavior near the origin consistent with the limit in Part 3.

Laboratory Exercise 2.3

Applications of Limits

Name ______________________ Due Date __________

Solve each of the following problems and explain how you make use of limits.

1. If you invest \$5000 in an account that pays 6% interest compounded n times each year, then after 20 years you will have $5000\left(1+\frac{0.06}{n}\right)^{20n}$ dollars in the account.

   (a) How much money will you have if the interest is compounded quarterly?

   (b) How much money will you have if the interest is compounded monthly?

   (c) How much money will you have if the interest is compounded *continuously*? (Continuous compounding is the limit as the number of compounding periods goes to infinity.)

2. The number of animals as a function of time in a certain population is given by

$$P(t) = \frac{300}{1 + 7(0.69)^t}$$

What is the maximum number of animals the environment can support?

3. A rocket ship is launched from the surface of the Earth. As it burns fuel, its mass decreases. When the rocket ship is a distance d from the center of the Earth, its mass is given by

$$\text{Mass} = 50000\left(1 + 9(4.3)^{10^7\left(\frac{1}{d} - \frac{1}{R}\right)}\right) \text{ kilograms}$$

where $R = 6.38 \times 10^6$ meters is the radius of the Earth. What is the initial mass of the rocket ship? What percentage of the original mass is fuel?

Laboratory Exercise 2.4

One-Sided Limits

Name ______________________________ **Due Date** ____________

Estimate the left-hand and right-hand limits as $x \to 0$ of each of the following functions numerically and graphically. Explain your answers and check them by using *DERIVE* to calculate the limits. Determine if the (two-sided) limit exists and explain your answers.

1. $f(x) = \dfrac{\sin |x|}{x}$

2. $g(x) = x \left| \sin \left(\dfrac{1}{x} \right) \right|$

3. $h(x) = \begin{cases} \frac{4^x - 1}{2x} \text{ if, } & x < 0 \\ \frac{2^x - 1}{x} \text{ if, } & x > 0 \end{cases}$

4. $k(x) = \dfrac{1 - \cos(3x)}{x}$

Laboratory Exercise 2.5

Asymptotes

Name ______________________________ Due Date ____________

Plot the graphs of each of the following functions and find their horizontal and vertical asymptotes. Explain how you got your answers. Include the graphs of the asymptotes in the picture.

1. $f(x) = \dfrac{5x^3 - 1}{2x^3 + 5x^2 + 5x + 3}$

2. $g(x) = \dfrac{x^2 + 1}{6x^3 - 67x^2 + 112x + 45}$

3. $h(x) = \dfrac{x^3 + 4}{x^2 + x - 1}$

4. $k(x) = \dfrac{2x + 1}{x^4 + x^3 + x^2 + x + 1}$

Laboratory Exercise 2.6

A Phantom Asymptote

Name ______________________________ **Due Date** ____________

Let $f(x) = \dfrac{3x^5 - 5x^4 + 3x - 5}{3x^5 + x^4 - 13x^3 - x^2 + 13x - 5}$.

1. Find the zeros of the denominator of f.

2. Plot the graph of f.

3. Do all the zeros you found in Part 1 appear as vertical asymptotes in the graph from Part 2? If some are missing, explain what happened. (Hint: Factor the numerator.)

4. Explain what the graph of f looks like near each of the zeros you found in Part 1.

Laboratory Exercise 2.7

Illustrating the Definition of a Limit

Name ______________________ **Due Date** __________

1. Use *DERIVE* to verify that $\lim_{x \to 2\pi} \dfrac{(x - 2\pi)\sin x}{1 - \cos x} = 2$. Find a value of δ so that if $0 < |x - 2\pi| < \delta$, then $\left|\dfrac{(x - 2\pi)\sin x}{1 - \cos x} - 2\right| < 0.001$. Include appropriate graphs and explain your choice of δ.

2. Use *DERIVE* to verify that $\lim_{x \to \infty} \sqrt{x^2 + 3x + 5} - x = \frac{3}{2}$. Find a value of K so that if $x > K$, then $\left| \sqrt{x^2 + 3x + 5} - x - \frac{3}{2} \right| < 0.001$. Include appropriate graphs and explain your choice of K.

Chapter 3
Differentiation

New *DERIVE* topics •Calculating derivatives •Substituting in expressions •Loading and saving files •Using arbitrary functions •Saving and recalling files •Calculating implicit derivatives •Using the RHS function •Calculating higher order derivatives
Calculus concepts •Difference quotient •Derivative •Secant line •Tangent line •Implicit differentiation •High order derivatives

Solved Problem 3.1: The derivative and the difference quotient

New *DERIVE* Lessons •Calculating derivatives •Substituting in expressions

Let $f(x) = \cos x$.

(a) Use the difference quotient to estimate the value of $f'(2)$.

(b) Use the graph of the difference quotient to estimate the value of $f'(2)$.

(c) Find $f'(2)$ by asking *DERIVE* to calculate the appropriate limit.

(d) Find $f'(2)$ by asking *DERIVE* to calculate the derivative directly.

Solution to (a): According to the definition of derivative, $f'(2) = \lim_{h \to 0} \frac{\cos(2+h) - \cos 2}{h}$. Thus, we can estimate $f'(2)$ by plugging small values of h into the difference quotient. It will be convenient to define the difference quotient as a function. **Author** `D(h):=(COS(2+h)-COS(2))/h` as seen in expression 1 of Figure 3.1. Next we **Author** and **approX** the list [`D(−0.02)`, `D(−0.01)`, `D(0.01)`, Notice that we have used both positive and negative values of h. From expression 3 we conclude that $f'(2)$ is somewhere between −0.907203 and −0.911362. (How could you get a more accurate estimate?)

Solution to (b): To get the graph, highlight expression 1 and **Plot Beside**. We emphasize that the graph in window 4 of Figure 3.1 is the graph of the difference quotient as a function of h, not the graph of $\cos x$. To estimate the value of the derivative, we need to look at the difference quotient near $h = 0$. Thus we use the arrow keys to move the graphing cross near the point where the graph appears to cross the y-axis, **Center** and zoom in twice using F9. (Does

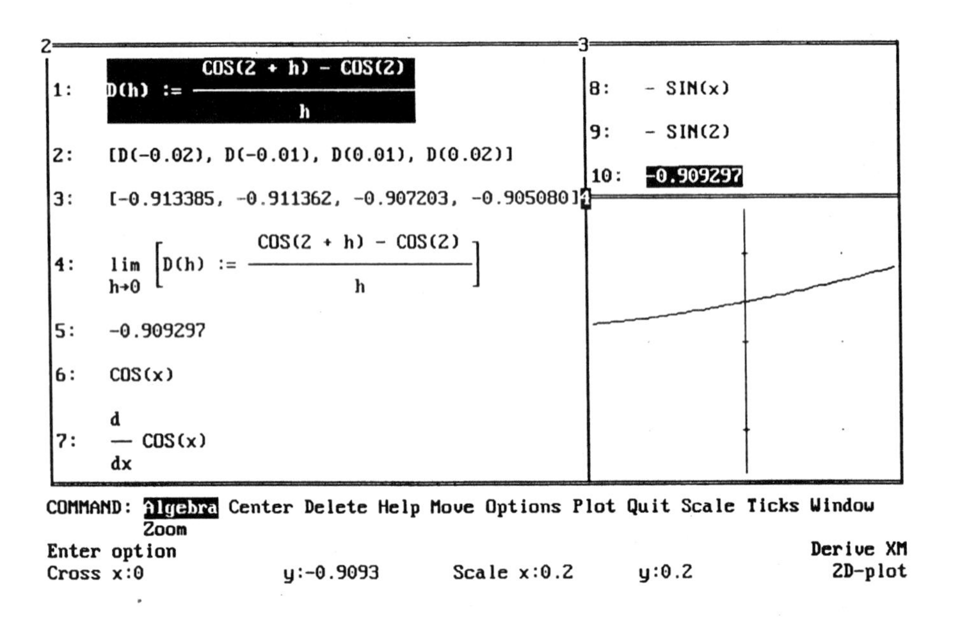

Figure 3.1: **The difference quotient and the derivative**

the graph actually intersect the y-axis?) From the lower left corner of Figure 3.1 we read the location of the graphing cross as x:0 y:-0.9093. This agrees with our estimate in Part (a), and we conclude that $f'(2) \approx -0.9093$.

Solution to (c): We need to calculate the limit of the difference quotient as h goes to zero. Thus we highlight expression 1 and use **Calculus Limit** setting Limit Point: 0. The limit appears in expression 4. We **approX** (rather than **Simplify**) to get expression 5. This is the correct value of the limit, and we note that our estimates in Parts (a) and (b) were indeed reasonably close to it.

Solution to (d): To have *DERIVE* calculate the derivative directly, we **Author** `COS x` as seen in expression 6 and then use **Calculus Differentiate** accepting all the default settings by pressing [Enter] at each prompt. (Press [C] for **Calculus** and you will be presented with a new menu, one of whose items is **Differentiate**. Press [D] and then [Enter] three times to accept the defaults.) The notation for the derivative now appears in expression 7, and we **Simplify** to get the answer in expression 8.

We want this evaluated at $x = 2$. Use **Manage Substitute** and type 2 at the prompt MANAGE SUBSTITUTE: value: x. The result, with x replaced by 2, appears in expression 9. We **approX** to get the decimal value in expression 10. We note that this agrees exactly with our answer in Part (c).

Solved Problem 3.2: Secant lines and tangent lines; local linearity

Let $f(x) = x\sqrt{1.4 - x}$.

(a) Plot the graph of f along with the graph of the secant line through $(1, f(1))$ and $(1.1, f(1.1))$.

(b) Include the graph of the secant line through $(1, f(1))$ and $(0.9, f(0.9))$.

(c) Include the graph of the tangent line through $(1, f(1))$.

(d) **Center** at $(1, f(1))$ and zoom in several times. Explain what you observe.

Solution to (a): First **Author** `F(x):=x SQRT(1.4-x)` and **Plot Beside**. In window 2 of Figure 3.2 we have used F9 to zoom in once. The slope of the secant line is given by the difference quotient, so in expression 2 we have defined $D(h)$ to be the difference quotient at $x = 1$. The slope of the secant line we want is $D(0.1)$, and it passes through the point $(1, f(1))$. Thus its equation is $D(0.1)(x-1)+f(1)$. This is expression 3, which we **Plot** to get the picture of the secant line. (Which one of the lines in window 2 is it?)

Solution to (b): The slope of this secant line is $D(-0.1)$, so its equation is $D(-0.1)(x-1) + f(1)$. This appears in expression 4 of Figure 3.2, which we **Plot** to get the picture in window 2.

Solution to (c): The slope of the tangent line is the derivative at $x = 1$. Thus we highlight expression 1 and use **Calculus Differentiate** and **Simplify** to get the derivative in expression 6. We next use **Manage Substitute** and type 1 at the prompt MANAGE SUBSTITUTE variable: x. The result, with x replaced by 1, is in expression 7. If we **approX** this (not shown in Figure 3.2) we obtain -0.158113. Thus the equation of the tangent line appears in expression 9 as $-0.158113(x - 1) + f(1)$. We **Plot** this to get the graph of the tangent line in window 2. (The tangent line is between the two secant lines.)

Solution to (d): In window 3 we see just the function (expression 1) and its tangent line (expression 9). To get this picture, we used F9 to zoom in until the **Scale** became x:0.02 y:0.02.

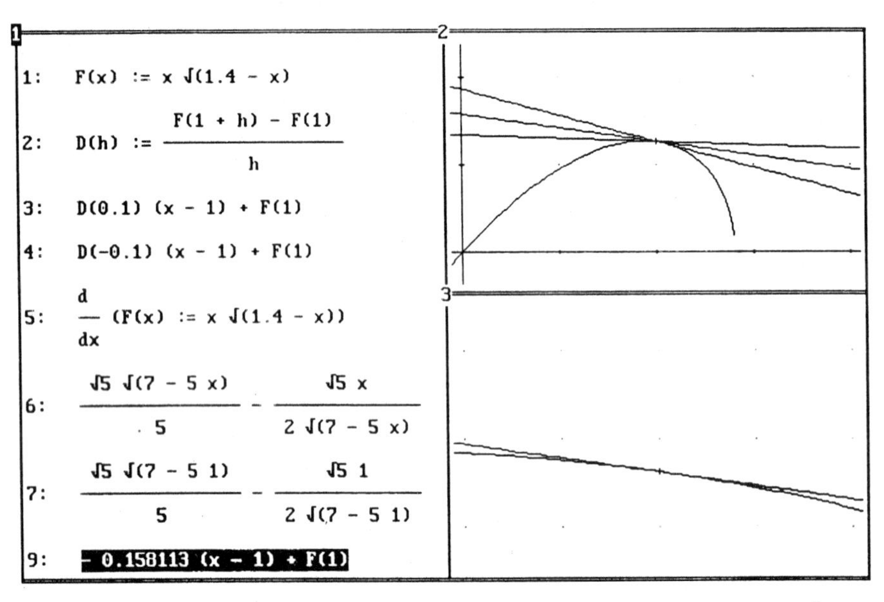

Figure 3.2: **Secant lines and tangent line for** $x(1.4 - x)^{\frac{1}{2}}$

We observe that with this scale, the graph of the function has almost merged with the graph of the tangent line. If you zoom in further, you will observe that the two graphs become indistinguishable. This illustrates the principle of *local linearity*: If you zoom in near enough on the graph of any function at a point where the derivative exists, its graph will be indistinguishable from the tangent line.

A Demonstration Using the SECANT.MTH File

New *DERIVE* Lessons •Loading *DERIVE* files

In Appendix II we have provided several *DERIVE* files that are designed to produce graphics. One of these is the SECANT.MTH file which shows secant lines converging to tangency. We will assume that the file has been saved following the instructions in Appendix II. To load it use **Transfer Merge** SECANT.

Now, let's use the file to show secant lines for $\sin x$ at $x = 1$. We **Author** `F(x):=SIN x` as seen in expression 6 of Figure 3.3. The function SECANT(a, n, s) makes a picture of n secant lines through the points $(a, f(a))$ and $(a + ks, f(a + ks))$ as k ranges from 1 to n in steps of s units along the x-axis. We will make four secant lines in steps of 0.2 along the x-axis. Thus we **Author** `SECANT(1, 4, 0.2)` and **approX**. We **Plot** the result in expression 8 to see the graph in window 2. If you want to get secant lines to the left of the point of interest, you just use a

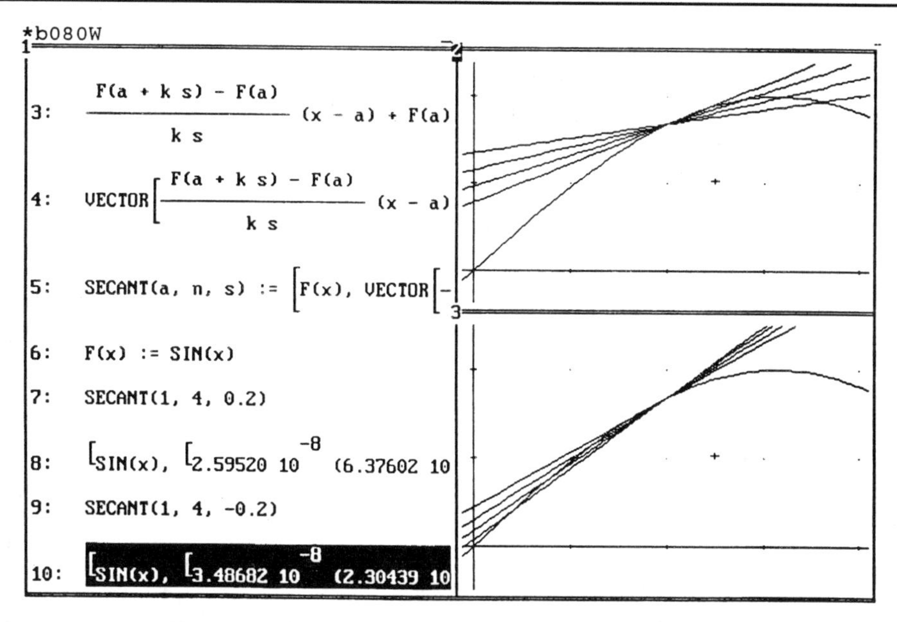

Figure 3.3: **Using the SECANT.MTH file**

negative value for s. In expression 9 we have used $s = -0.2$. The corresponding display is in window 3.

Solved Problem 3.3: Implicit differentiation

New *DERIVE* Lessons •Calculating implicit derivatives •Saving and recalling files •Using the RHS function

Let y be implicitly defined as a function of x by the equation $x^2y^3 - 2y = x^2 - 2y^2$ and the initial condition $y(1) = \dfrac{\sqrt{5}-3}{2}$.

(a) Use implicit differentiation to find y' in general and at the point $(1, \frac{\sqrt{5}-3}{2})$.

(b) Solve the given equation to determine y explicitly as a function of x.

(c) Calculate y' using the equation from Part (b) and show that you get the same answer as in Part (a).

Solution to (a): First **Author** `x^2y^3-2y=x^2-2y^2` as seen in expression 1 of Figure 3.4. A trick is required to get *DERIVE* to calculate implicit derivatives. It is necessary to discuss

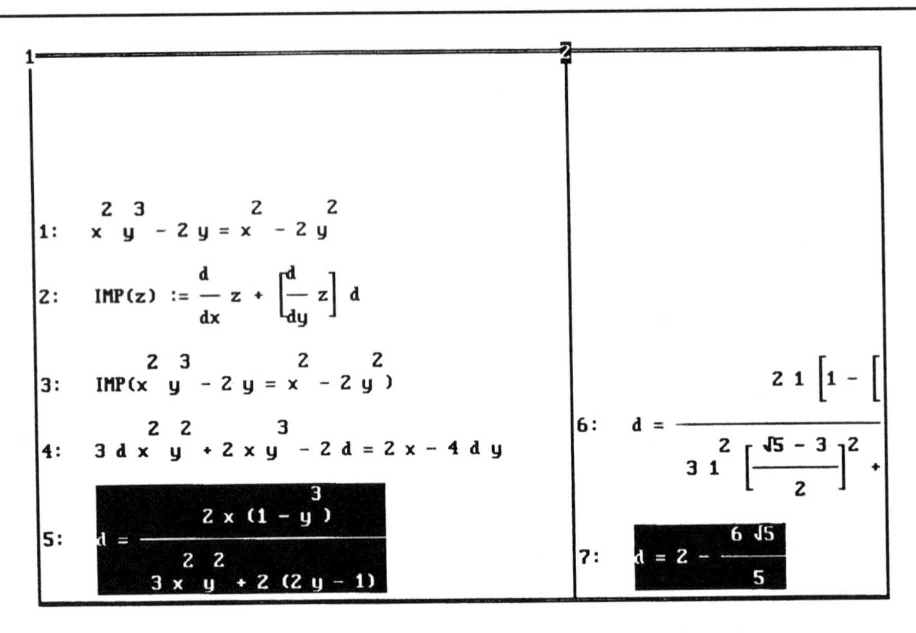

Figure 3.4: **Calculating an implicit derivative**

partial derivatives in order to explain why the trick works, but for now it is enough to know how to use it. **Author** `IMP(z):=dif(z,x)+dif(z,y)d`. Now if z is an equation in x and y and you **Author** and **Simplify** `IMP(z)`, *DERIVE* will return the implicitly differentiated equation with the letter d playing the role of y'. You may wish to save this function for future use with **Transfer Save Derive** IMP. You can then recall the function when you need it, using **Transfer Merge** IMP.

Author `IMP(#1)` and **Simplify**. The result appears in expression 4 of Figure 3.4. Now **soLve** expression 4, but when prompted with SOLVE variable: x type `d` to tell *DERIVE* that you want to solve for d rather than x. We see the implicit derivative in expression 5.

To find y' at the point $(1, \frac{\sqrt{5}-3}{2})$ we use **Manage Substitute** to replace x by `1` and y by `(SQRT(5)-3)/2`. **Simplify** to get the answer in expression 7.

Solution to (b): Use the arrow keys to highlight expression 1. **soLve** and type **y** in place of x at the prompt Solve variable: x. You will see the expressions 8 (not shown), 9, and 10 of Figure 3.5. An equation alone may allow several different definitions of y as a function of x as we see here, but you can check that only expression 10 has the property that $y(1) = \dfrac{\sqrt{5}-3}{2}$, so this is

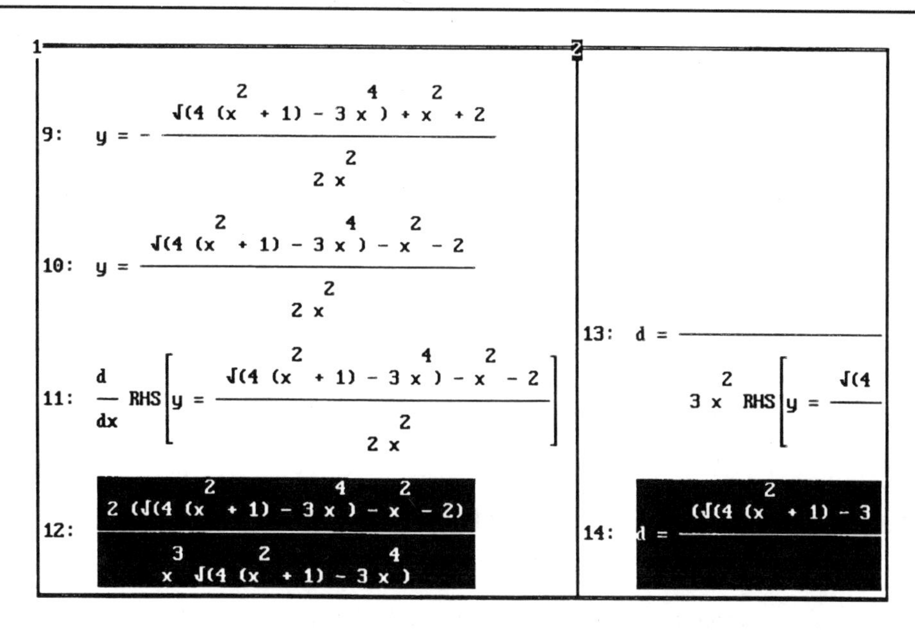

Figure 3.5: **Differentiating explicitly**

the function we are looking for.

Solution to (c): To get the explicit derivative, highlight expression 10 of Figure 3.5 and use **Calculus Differentiate**, but at the first prompt expression: #10, delete what is there and type `RHS(#10)`. (This tells *DERIVE* that we want to use only the right-hand side of expression 10.) Complete the entry as usual and **Simplify** to get the derivative partially displayed in expression 12.

To show that our derivatives are the same, we need to replace y in expression 6 of Figure 3.4 by expression 10 of Figure 3.5. Highlight expression 5 and use **Manage Substitute**. When prompted with MANAGE SUBSTITUTE value: x press Enter to indicate that x is to be left as it is. When prompted with MANAGE SUBSTITUTE value: y type `RHS(#10)`. **Simplify** to get the large expression that is partially displayed in expression 14 of Figure 3.5. To show that expression 12 is the same as the right-hand side of expression 13 **Author** `#12-RHS(#14)` and **Simplify**. *DERIVE* will return 0.

A Note on Calculating Higher Order Derivatives

New *DERIVE* Lessons •Calculating higher order derivatives

Suppose you wish to calculate the fourth derivative of $\cos x$. Use **Calculus Differentiate**, but when you see the prompt CALCULUS DIFFERENTIATE: Order: 1, type **4** in place of 1.

You will see that *DERIVE* uses the notation $\left[\frac{d}{dx}\right]^4$ in place of the more familiar $\frac{d^4}{dx^4}$, but the meaning is the same. **Simplify** to get the answer.

Practice Problems

1. Plot the graph of $f(x) = \cos x$. Include the graph of the secant lines through $(1, f(1))$ and $(2, f(2))$ and through $(0.5, f(0.5))$ and $(1, f(1))$ as well as the graph of the tangent line at $x = 1$.

2. Use the SECANT.MTH field to produce a picture of $f(x) = \dfrac{1}{1+x^2}$ together with at least 6 of its secant lines at $x = 1$.

3. Let $f(x) = \dfrac{x}{x^2+1}$. Evaluate the difference quotient at $x = 3$ and $h = 0.01$.
   Answer: -0.0798204

4. Using the function f from Problem 3, calculate $f'(3)$ by finding the limit as h goes to zero of the difference quotient. Check your answer by asking *DERIVE* to calculate the derivative directly. Answer: -0.08

5. If $s(t) = \sin\sqrt{t}$ gives position, find the average velocity from $t = 1$ to $t = 1.5$. Answer: 0.0661655. Find the instantaneous velocity at $t = 1$. Answer: 0.270151

6. Let $f(x) = \sqrt{1+x^2}$. Find the difference quotient at $x = 2$. Answer: $\dfrac{\sqrt{h^2+4h+5}-\sqrt{5}}{h}$.
   Calculate its limit as h goes to 0. Answer: $\dfrac{2}{\sqrt{5}}$

7. Use *DERIVE* to calculate the derivative of y with respect to x for the function defined implicitly by $\sin(xy) = \cos(x+y)$. Answer: $-\dfrac{\sin(x+y)+y\cos(xy)}{\sin(x+y)+x\cos(xy)}$

8. Let y be the (real-valued) function defined by $y^3 + x = y + xy$, $y(0) = 0$. Find y. Answer:
   $y = \dfrac{\sqrt{4x+1}-1}{2}$

9. If $f(x) = \cos(x^2)$, find $f^{(7)}(1)$. Answer: -2538.09

Laboratory Exercise 3.1

Estimating the Derivative Using the Difference Quotient

Name ______________________ Due Date __________

Let $f(x) = \tan(x^2)$.

1. Estimate $f'(3)$ by plugging small values of h into the difference quotient.

2. Estimate $f'(3)$ using the graph of the difference quotient. Explain how you got your answer.

3. Find $f'(3)$ by calculating the limit of the difference quotient.

4. Use *DERIVE* to find $f'(3)$ directly.

Laboratory Exercise 3.2

Average Velocity of a Car

Name ______________________________ **Due Date** __________

For each of the following problems explain how you got your answers, and be sure to include units.

1. You drive for one minute at a constant velocity of 55 miles per hour. You then instantly slow down and drive for one more minute at a constant velocity of 30 miles per hour. What is your average velocity for the two minutes?

2. You drive for one mile at a constant velocity of 55 miles per hour. You then instantly slow down and drive for one more mile at a constant velocity of 30 miles per hour. What is your average velocity for the two miles?

3. You drive for one minute at a constant velocity of 30 miles per hour. You then want to instantly speed up and drive another minute so that your average velocity for the two minutes is 60 miles per hour. What velocity must you drive for the second minute?

4. You drive for one mile at a constant velocity of 30 miles per hour. You then want to instantly speed up and drive another mile so that your average velocity for the two miles is 60 miles per hour. What velocity must you drive for the second mile? (Be careful!)

Laboratory Exercise 3.3

Sky Diving

Name ____________________ Due Date __________

A sky diver jumps from an airplane. During the period before her parachute opens, she falls $986(0.835^t - 1) + 176t$ feet in t seconds.

1. What is her average velocity over the time periods $t = 1.99$ to $t = 2$ and $t = 2$ to $t = 2.01$?

2. Use your answer in Part 1 to estimate the instantaneous velocity at $t = 2$. Explain how you got your answer.

3. Use *DERIVE* to find the exact value of the instantaneous velocity and compare with your answer in Part 2. (Note: When you differentiate, your answer will involve functions and constants that you may not be familiar with. Just substitute in the appropriate value for t and **approX**.)

4. Assume that the sky diver opens her parachute one minute after jumping from the airplane and lands on the ground 30 seconds later. Graph the distance she falls as a function of time over the first minute of the fall.

5. At $t = 60$ seconds, the parachute opens, slowing the sky diver's fall. We will not give you a formula for the distance fallen from $t = 60$ to $t = 90$. Using pencil and paper, extend the graph in Part 3 to make a graph that would reasonably describe distance fallen as a function of time over the entire period of the fall.

Laboratory Exercise 3.4

Secant Lines and Tangent Lines

Name ______________________ Due Date __________

Let $f(x) = \dfrac{6x}{2 + x^2}$.

1. Plot the graph of $f(x)$.

2. Find the equation of the secant lines through $(1, f(1))$ and $(2, f(2))$ and through $(1, f(1))$ and $(1.5, f(1.5))$. Include these graphs in your picture from Part 1.

3. Find the equation of the tangent line to the graph of f at $x = 1$ and include its graph in your picture.

4. Explain the relationships among the secant lines and the tangent line.

5. Use the **Delete** command to delete all the graphs except the graph of f. **Center** at the point $(1, f(1))$ and zoom in until the graph looks like a straight line. What scale is necessary to accomplish this?

6. Explain how you can use your graph in Part 5 to estimate the derivative at $x = 1$. (Hint: Use the arrow keys to move the cross to two different points on the "line.")

Laboratory Exercise 3.5

The Derivative of the Gamma Function

Name ______________________________ **Due Date** ____________

DERIVE knows a function called the *Gamma function*. To enter it, **Author** Alt G **x**. You should see on the screen $\Gamma(x)$. This function is very important in mathematics, and you may encounter it later when you study *improper integrals*. (Basically the Gamma function fills in the gaps in the graph of $(n-1)!$) We use the Gamma function here because of its importance, and because it is a function whose derivative is difficult to calculate explicitly.

1. Simplify $\Gamma(1)$, $\Gamma(2)$, $\Gamma(3)$, $\Gamma(4)$, and $\Gamma(5)$. Compare the answers with $(n-1)!$

2. Try to find the derivative of $\Gamma(x)$ at $x = 1$ by asking *DERIVE* to calculate the limit of the difference quotient. Does the answer make sense? (You might also try **Calculus Differentiate**, but you won't have any better luck.)

   DERIVE failed because it doesn't know how to find the limit any more than you do. In the remainder of this exercise, we will use several strategies to estimate the derivative.

3. Estimate the derivative by approximating the difference quotient using small values of h.

4. Plot the graph of the difference quotient as a function of h, and use this graph to estimate the derivative at $x = 1$. Compare your answer here with the one you got in Part 3.

5. Plot the graph of $\Gamma(x)$. (Suggestion: This will be significantly faster if you move the graphing cross to $(2, 2)$ and **Center** before plotting.)

6. **Center** and zoom in on the point $(1, \Gamma(1))$ until the graph looks like a straight line. Use this "line" to approximate the derivative and compare it with your previous answers. (Hint: See the hint in the previous laboratory exercise.)

7. Find the equation of the tangent line to $\Gamma(x)$ at $x = 1$ using one of your previous estimates for the derivative.

8. Plot the graphs of $\Gamma(x)$ and of the tangent line in the same window. Zoom in on the point of tangency and explain how these graphs support your estimate of $\Gamma'(1)$.

Laboratory Exercise 3.6

Existence of Derivatives

Name ______________________________ **Due Date** ____________

Recall that a function fails to have a derivative where it is discontinuous or where its graph is not "smooth."

1. Plot the graph of $|x^2 - 5x + 6|$.

2. Find all points where the derivative fails to exist. Explain your answers.

3. Plot the graph of $\frac{|x^4 - x - 1| - x}{x^2 + x - 1}$.

4. Find all points where the derivative fails to exist. Explain your answers.

Laboratory Exercise 3.7

Implicit Derivatives

Name ______________________________ **Due Date** ____________

Let y be implicitly defined as a function of x by the equation $x^2y^4 - 2y^3 = yx^2 - 2y^2$ and by the initial condition $y\left(\frac{1}{2}\right) = \dfrac{7-3\sqrt{5}}{2}$.

1. Show that the point $(\frac{1}{2}, \frac{7-3\sqrt{5}}{2})$ satisfies the equation $x^2y^4 - 2y^3 = yx^2 - 2y^2$.

2. Use implicit differentiation to find the derivative of y with respect to x.

3. Find the derivative of y with respect to x at the point $(\frac{1}{2}, \frac{7-3\sqrt{5}}{2})$.

4. Solve the given equation to determine y explicitly as a function of x.

5. Use your answer in Part 4 to calculate y' explicitly.

6. Show that your answers in Parts 2 and 5 are the same.

Laboratory Exercise 3.8

Higher Order Derivatives

Name ______________________________ **Due Date** ____________

For each of the following functions, use *DERIVE* to calculate derivatives of orders 1, 2, 3, 4, and 5. Using your answers, conjecture a formula for the nth derivative in each case.

1. $\sin x$

2. x^{17}

3. $\dfrac{1}{1-x}$

4. $xf(x)$ where $f(x)$ is an arbitrary function. (Suggestion: **Author** `F(x):=`, and then look at derivatives of `xF(x)`.)

Chapter 4

Applications of the Derivative

New *DERIVE* topics
Calculus concepts •Intervals of increase and decrease •Local maxima and minima •Absolute maxima and minima •Concavity •Points of inflection

Solved Problem 4.1: Increasing and decreasing functions

Plot the graphs of $f(x) = 2\sqrt{x^4+1} - 2x^2 + x$ and its first derivative. Use the first derivative to find all intervals where f is increasing.

Solution: **Author** 2SQRT(x^4+1)-2x^2+x and **Plot** to get the graph in window 2 of Figure 4.1. Next use **Calculus Differentiate** and **Simplify** to get the derivative in expression 3. Its graph is in window 3 of Figure 4.1. f is increasing where the first derivative is positive; that

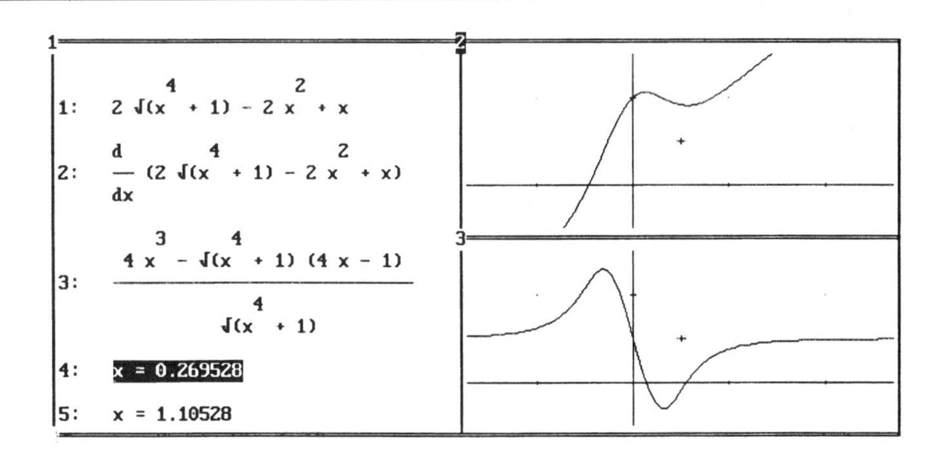

Figure 4.1: **Graphs of** $2(x^4+1)^{\frac{1}{2}} - 2x^2 + x$ **and its derivative**

is where the graph in window 3 is above the x-axis. From window 3 we see that is where the graph of f' crosses the x-axis between 0 and 1 and again between 1 and 2. We change to the approximate mode using **Options Precision Approximate** Enter and **soLve** expression 3 with Upper:0 Lower:1. The result is in expression 4 of Figure 4.1. When we **soLve** again on the interval Lower:1 Upper:2, we get expression 5. From the location of the zeros and the graph in window 3, we conclude that the first derivative is positive on the intervals $(-\infty, 0.269528)$ and $(1.10528, \infty)$, and hence these are the intervals on which f is increasing. The graph of f in window 2 supports this conclusion.

A Demonstration for Increasing and Decreasing Functions: The INCREASE.MTH file in Appendix II makes a graphical demonstration if you have a color monitor. We will assume that you have entered and saved the file according to the instructions in Appendix II. Load it using **Transfer Merge** INCREASE. The next step is to tell *DERIVE* which function we want to use. For example, if we want to look at $\cos x$, we **Author** `F(x):=COSx`. Now highlight the expression that begins with `plot_this` and **Plot**. *DERIVE* will plot the graph of $F(x)$ in one color on intervals where it is increasing and another color on intervals where it is decreasing. There is a file, CONCAVE.MTH in Appendix II, which can produce a similar demonstration for concavity.

Solved Problem 4.2: Finding maxima and minima

(a) Plot the graph of $f(x) = \dfrac{2x^2 + x - 1}{2x^2 + 5x + 4}$ and estimate the local extrema by zooming in on appropriate points. Then use the derivative to calculate the extrema.
(b) Find all local maxima and minima of $|x^2 - \cos x| - x$. Identify any global maxima and minima.

Solution to (a): First **Author** `(2x^2+x-1)/(2x^2+5x+4)` and **Plot**. From the graph in window 3 of Figure 4.2 we see that this function has a single local maximum and a single local minimum. In window 4 we have moved the graphing cross to the minimum, **Center**ed, and zoomed in twice using F9 to get a closer look. From the lower left corner of the screen, we read the location of the cross as x: -0.5972 y: -0.5125 , and thus the minimum is located at approximately $(-0.5972, -0.5125)$. A similar analysis shows that the maximum is at approximately $(-1.9138, 2.5125)$. Use **Calculus Differentiate** and **Simplify** to get the derivative as seen in expression 3 of Figure 4.3. When we **soLve** this, *DERIVE* gives the two critical points in expressions 4 and 5. (Some versions of *DERIVE* will also report the solution $\frac{1}{0}$ which should be ignored.) Highlight expression 4 and **approX** to get expression 7. This is the x-coordinate of the local maximum we approximated in (a). We want to plug the right-hand side of this equation into expression 1. Highlight expression 1 and use **Manage Substitute**. At the prompt MANAGE SUBSTITUTE value:x, type `RHS(#7)`, and **approX** to get expression 9. This gives the location of the maximum as $(-1.91143, 2.51185)$, and this compares favorably with our earlier estimate. We **approX** expression 5 to get expression 10, and from expression 12 we locate the minimum at $(-0.588562, -0.511858)$.

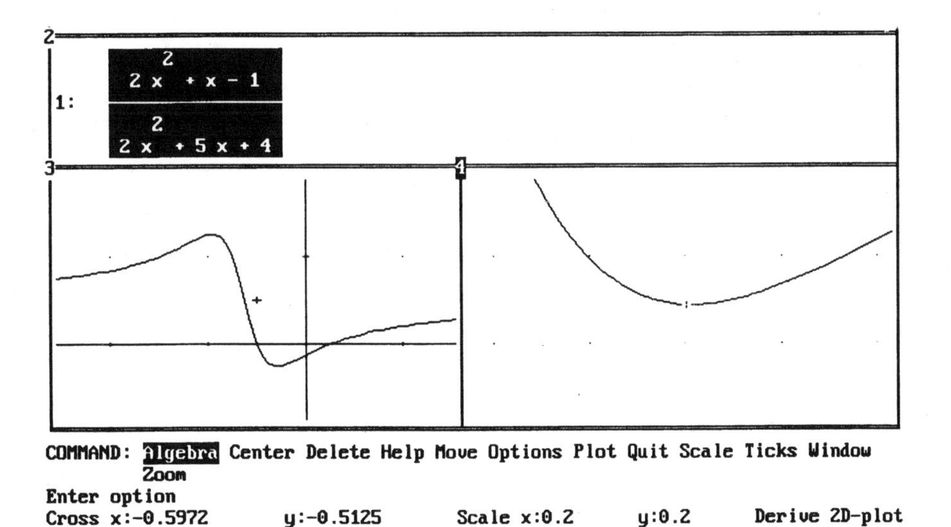

Figure 4.2: **Zooming in on a minimum**

1: $\dfrac{2x^2 + x - 1}{2x^2 + 5x + 4}$

2: $\dfrac{d}{dx} \dfrac{2x^2 + x - 1}{2x^2 + 5x + 4}$

3: $\dfrac{8x^2 + 20x + 9}{(2x^2 + 5x + 4)^2}$

4: $x = -\dfrac{\sqrt{7}}{4} - \dfrac{5}{4}$

5: $x = \dfrac{\sqrt{7}}{4} - \dfrac{5}{4}$

7: x = -1.91143

8: $\dfrac{2\ \mathrm{RHS}(x = -1.91143)^2 + \mathrm{RHS}(x = -1.91143)}{2\ \mathrm{RHS}(x = -1.91143)^2 + 5\ \mathrm{RHS}(x = -1.91143)}$

9: 2.51185

10: x = -0.588562

11: $\dfrac{2\ \mathrm{RHS}(x = -0.588562)^2 + \mathrm{RHS}(x = -0.588562}{2\ \mathrm{RHS}(x = -0.588562)^2 + 5\ \mathrm{RHS}(x = -0.58856}$

12: -0.511857

Figure 4.3: **Finding critical points**

Solution to (b): **Author** `|x^2-COS x|-x` and **Plot**. From the graph in window 2 of Figure 4.4 we see that there are two local minima and a single local maximum. There is no global maximum, but the minimum in the fourth quadrant is a global minimum. Notice that at the maximum, the graph is smooth, and we expect this to correspond to a point where the first derivative is zero, but at both minima the graph is not smooth, indicating that the derivative does not exist.

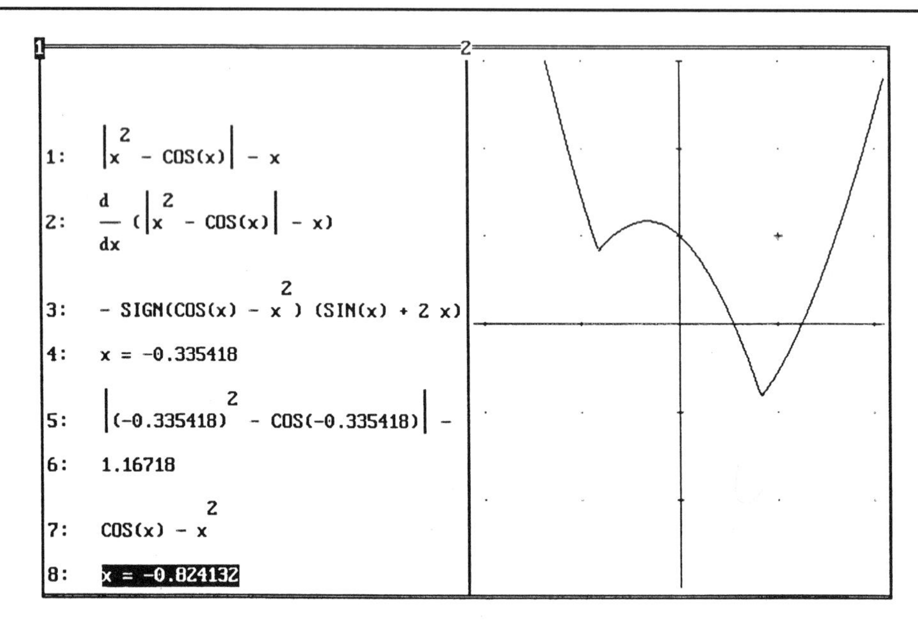

Figure 4.4: **Extrema of** $|x^2 - \cos x| - x$

Use **Calculus Differentiate** and **Simplify** to get the derivative in expression 3. The local maximum corresponds to a zero of the first derivative which occurs between $-\frac{1}{2}$ and 0. We change to the approximate mode using **Options Precision Approximate** [Enter] and then **soLve** on the range Lower:`-1/2` Upper:`0`. The result appears in expression 4. We use **Manage Substitute** to plug this value into expression 1 and **approX** to get expression 6 which shows that the maximum is at $(-0.335418, 1.16718)$.

To find the minima, we note that the derivative involves the SIGN function which is 1 for positive x, -1 for negative x, and does not exist at 0. Thus the derivative is undefined when the argument, $x^2 - \cos x$, of the SIGN function is zero. From the graph in window 2 of Figure 4.4, we expect this to occur once on the interval $(-1, 0)$ and again on the interval $(0, 1)$. **Author** `COSx-x^2` and **soLve** on Lower:`-1` Upper:`0` to get the result in expression 8. Since this is a number that makes $x^2 - \cos x = 0$, the function value is 0.824132, and one minimum occurs at

$(-0.824132, 0.824132)$.

If we **soLve** expression 3 once more on the range `Lower:0` `Upper: 1`, we find that the absolute minimum occurs at $(0.824132, -0.824132)$.

Solved Problem 4.3: Points of inflection

Plot the graph of $f(x) = x^6 - x^4$ and its second derivative. Find all inflection points.

Solution: First **Author** `x^6-x^4` and use **Calculus Differentiate** setting the `Order:` field to 2 to get the second derivative in expression 3 of Figure 4.5. The graph of f, which shows two

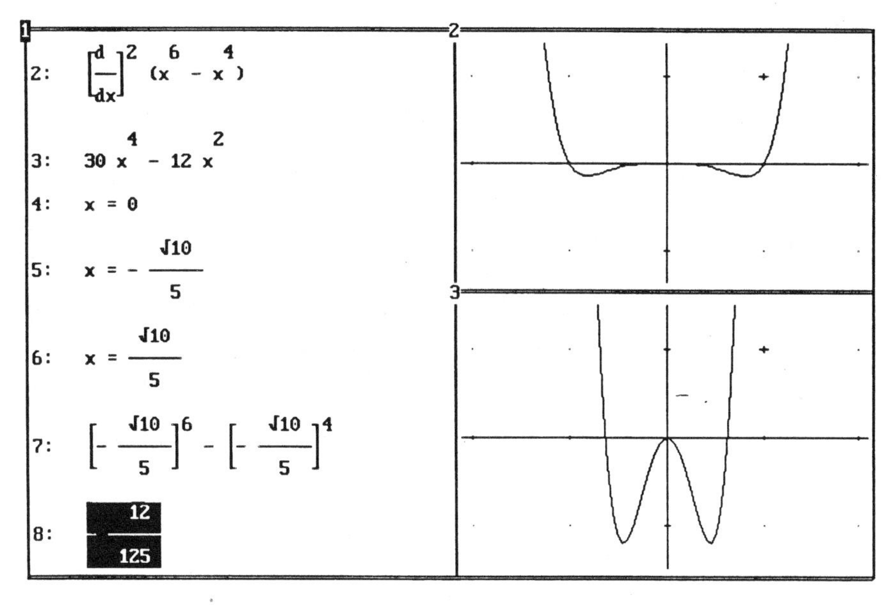

Figure 4.5: **Inflection points of $x^6 - x^4$**

points of inflection, is in window 2 and its second derivative in window 3. We **soLve** expression 3 to get the three zeros in expressions 4, 5, and 6. We note that $x = 0$ does not correspond to a point of inflection since the graph of the second derivative shows that it does not change sign. Highlight f and use **Manage Substitute** to plug in $-\dfrac{\sqrt{10}}{5}$. From expression 8 we see that one inflection point occurs at $\left(-\dfrac{\sqrt{10}}{5}, -\dfrac{12}{125}\right)$, and the symmetry of the graph ensures that the other occurs at $\left(\dfrac{\sqrt{10}}{5}, -\dfrac{12}{125}\right)$.

Solved Problem 4.4: Assembling a cardboard box

A cardboard box with a square base and no lid is to have a volume of 4 cubic feet. The box is to be assembled by taping the four seams around the bottom and one seam up the side. Cardboard costs 11 cents per square foot, and taping costs 8 cents per foot. Find the dimensions of the box that costs the least to build.

Solution: If x denotes the length of the base of the box and y the height, then there are x^2 square feet in the base of the box and four sides of area xy square feet. Thus, the cost of the cardboard expressed in dollars is $0.11(x^2 + 4xy)$. We must tape four seams of length x and one of length y which gives a taping cost of $0.08(4x + y)$. The total cost of production is $0.11(x^2 + 4xy) + 0.08(4x + y)$. The volume of the box is 4 cubic feet so that $x^2y = 4$, or $y = \dfrac{4}{x^2}$. Plugging this into the cost formula for the box, we obtain the cost $0.11\left(x^2 + 4x\dfrac{4}{x^2}\right) + 0.08\left(4x + \dfrac{4}{x^2}\right)$, which appears as expression 1 of Figure 4.6. Its graph in window 2 shows that the minimum cost occurs somewhere between $x = 1$ and $x = 2$. Use **Calculus Differentiate** and **Simplify** to get the derivative in expression 3. Its graph in window 3 and verifies that the critical point occurs between $x = 1$ and $x = 2$. We change to the approximate mode using **Options Precision Approximate** and **soLve** on this interval. The result, the length of the base, is in expression 4. Using the relationship $y = \dfrac{4}{x^2}$, we have found the height of the box of minimum cost in expression 6.

Solved Problem 4.5: Newton's method

Apply five steps of Newton's method to $f(x) = x + \cos x$ beginning at $x = -3$ to approximate a zero. Use the NEWTON.MTH file provided in Appendix II to make a picture of this procedure.

Solution: The derivative of f is $1 - \sin x$; thus the formula given by Newton's method is $x_{n+1} = x_n - \dfrac{x_n + \cos x_n}{1 - \sin x_n}$. **Author** `N(x):=x-(x+COS x)/(1-SIN x)` as seen in expression 1 of Figure 4.7. Newton's method says $x_{n+1} = N(x_n)$, so taking $x_0 = -3$, we can get x_1 if we **Author** and **approX** `N(-3)`. The result of one step of Newton's method is in expression 3. To

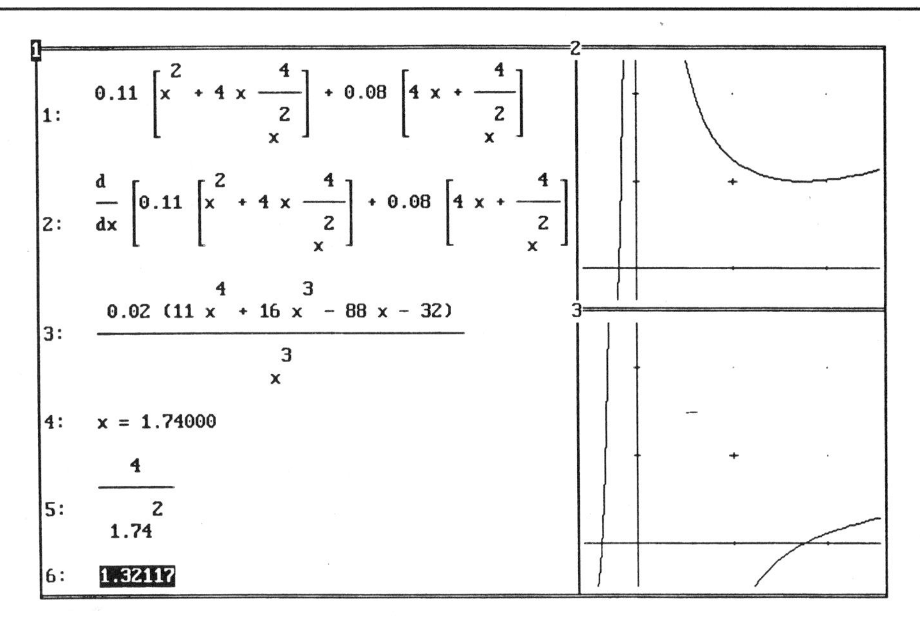

Figure 4.6: **A least cost box**

get the second step, we could **Author** and **approX** `N(0.496557)`, but there is an alternative method that saves a good deal of typing. The F4 key brings the highlighted expression to the author line enclosed in parentheses. Thus we **Author** `N` F4 Enter, and expression 4 appears. **approX** to get the result of two steps of Newton's method in expression 5. We continue this procedure until the method appears to be converging as in expressions 9 and 11. Expression 11 is accurate to six digits.

To make the picture, you must first enter and save the NEWTON.MTH file according to the instructions in Appendix II. Load it using **Transfer Merge** NEWTON. To tell *DERIVE* which function we are using, we **Author** `F(x):=x+COS x`. The function NPIC(a, k) makes a picture of k iterations of Newton's method applied to $F(x)$ with starting point $x = a$. Thus, we **Author** `NPIC(-3,5)` and **approX**. The result, partially displayed in expression 23 of Figure 4.7, is designed to tell *DERIVE* how to draw the picture. **Plot** to open a plot window and **Plot** again. *DERIVE* will present PLOT: Min: −3.1416 Max: 3.1416. You must change this to Min:`0` Max:`1`. (Use the Tab key to get to the Max: field.) Now press Ctrl Enter, and the picture in window 3 will appear.

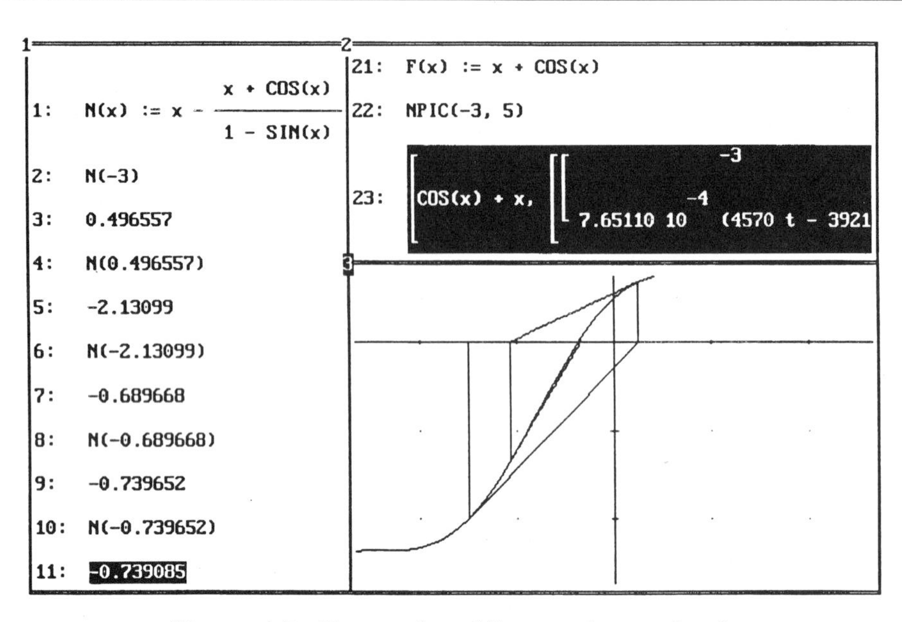

Figure 4.7: **Executing Newton's method**

Practice Problems

1. Plot the graph of $x^3 - x^2 + 1$ and use the graphing cross to locate the local maxima and minima. Answer: Maximum at $(0, 1)$, minimum at $(0.66, 0.85)$
2. Find the critical points of $x^4 + x^3 - 2$. Answer: 0 and $-\dfrac{3}{4}$
3. Use the INCREASE.MTH file to show where $f(x) = \dfrac{x}{1+x^2}$ is increasing and where it is decreasing.
4. Find the zeros of the second derivative of $x^3 - \cos x$. Answer: -0.164418
5. Apply five steps of Newton's method with starting point $x_0 = 2$ to approximate a solution of $\sin(\cos x) = \sin x$. Answer: -21.2520

Laboratory Exercise 4.1

Intervals of Increase, Decrease, and Concavity

Name ________________________________ **Due Date** ____________

Graph each of the following functions and their first derivatives. Identify the intervals where the function is increasing and those where the function is decreasing. Also plot the second derivative and find the intervals where the function is concave up and intervals where the function is concave down. Explain how you got your answers.

1. $\dfrac{2x^2 + x - 1}{2x^2 + 5x - 4}$

2. $\sqrt{x^3 + x + 1} - \sqrt{x^3 + 1}$

3. $\dfrac{x}{1+x^6}$

4. $\sqrt{x} - \sin x$

Laboratory Exercise 4.2

Extrema of Functions

Name ______________________ **Due Date** ____________

For each of the following functions

(a) Plot the graph and use the graphing cross to estimate all local maxima and minima.

(b) Identify any global maxima and minima.

(c) Use the derivative to find all local maxima and minima and explain your work.

(d) Find all points of inflection.

1. $x^4 + x^3 + x + 1$

2. $\dfrac{x^2 + x + 1}{x^4 + x + 1}$

3. $x^3 - \sin x$

4. $|x^2 - \cos x| - |x + \sin x - 1|$

Laboratory Exercise 4.3

A Population of Elk

Name ______________________________ **Due Date** ______________

A breeding group of elk is introduced into a protected region at time $t = 0$. t years later, the *logistic model of population growth* predicts that the population will be

$$P(t) = \frac{15.45}{0.03 + (0.55)^t}$$

1. How many elk were placed in the protected area? Explain how you got your answer.

2. Plot the graph of $P(t)$ and explain how the population varies with time.

3. The *maximum sustainable yield* model for wildlife management says that the population should be maintained at a level where population growth is a maximum. Express this statement in terms of inflection points of $P(t)$.

4. According to the maximum sustainable yield model, at what level should the elk population be maintained? Show your work.

5. How many elk can the environment support? Explain how you got your answer.

Laboratory Exercise 4.4

Applications of Extrema I

Name ____________________ Due Date __________

1. **Building boxes**: A contractor wants to bid on an order to make 200,000 boxes out of cardboard that costs 13 cents per square foot. The base of each box is square and must be reinforced with an extra piece of cardboard. The boxes must be assembled by taping the four seams around the bottom, one seam up the side, and one seam on top to make a hinged lid. Taping costs 9 cents per foot, and the boxes are to hold 5 cubic feet.

   (a) Write the cost of a single box as a function of the length of the base.

   (b) What should the dimensions of the box be to ensure lowest cost? Show your work.

   (c) If the contractor wishes to make a profit of 19%, how much should she bid?

   (d) A competitor, who also wants to make a profit of 19%, prepares a bid for boxes that are cubes. By how many dollars does this competitor lose the bid?

2. **A water tank**: A customer wants to know the cost of a water tank that will hold 800 gallons. Your firm can make the tank from a rectangular piece of metal that you will roll into a cylinder and two circular ends that will be welded to the cylinder. The rectangular piece is made of a malleable alloy that costs $2.36 per square foot, but the circular ends can be made of a cheaper metal costing $1.44 per square foot. Welding costs $1.20 per foot.

   (a) Write the cost of the tank as a function of the radius of the base. (There are 7.5 gallons in a cubic foot.)

   (b) What are the dimensions that ensure lowest cost? Show your work.

   (c) To cover shipping and make a small profit, you need to add 23% to the cost of building the tank. What price do you quote to the customer?

   (d) Surprised at the cost, the customer informs you that she has only $225 to spend. What is the tank of greatest volume you can sell her for that price? Don't forget about the 23% profit.

Laboratory Exercise 4.5

Application of Extrema II

Name ______________________ Due Date __________

1. **Building a road**: A road from city A to city B must cross a strip of private land and a strip of public land as shown in Figure 4.8. It costs \$73,700 per mile to build a road on public land, but it costs 35% more to build the road on private land. What is the minimum cost of the project? Draw a map with distances labeled that shows the road of minimum cost.

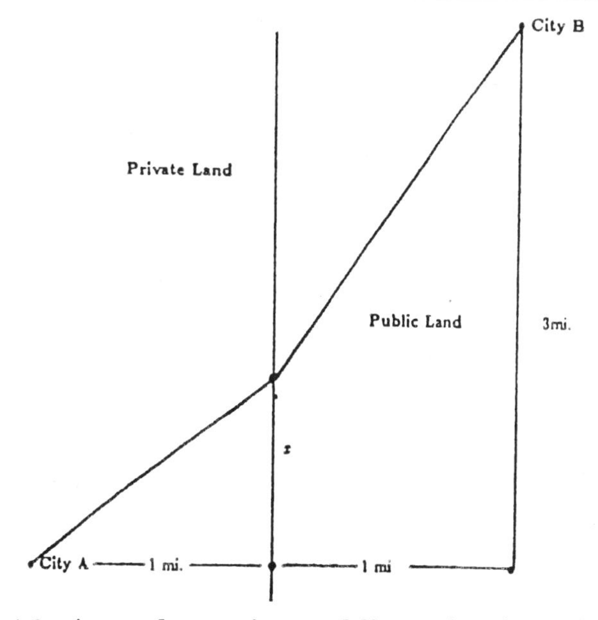

Figure 4.8: **A road crossing public and private land**

2. **An ill-fated boat**: Let the y-axis and the line $x = 1$ represent the banks of a river flowing in the negative y direction. A boat enters the river at the point $(1, 0)$ and maintains its heading toward the origin. The effort of the under powered boat to reach the origin in a strong current causes the boat to follow the graph of $f(x) = \dfrac{x^{3.33} - 1}{2x^{0.66}}$.

   (a) Plot the graph of the path taken by the boat and discuss its fate.

   (b) What is the nearest the boat comes to the origin? At what point does this occur? Explain how you arrived at your answer.

   (c) At the time the boat is nearest to the origin, its velocity in the y direction is $\dfrac{dy}{dt} = -5$ miles per hour. What is its velocity in the x direction?

Laboratory Exercise 4.6

Convergence of Newton's Method and Starting Points

Name ______________________________ **Due Date** ____________

1. Apply five steps of Newton's method to $\cos x + \sin x$ with starting point $x_0 = 0.6$ and use the NEWTON.MTH file to make a picture of the procedure.

2. Make a picture of 10 iterations of Newton's method with starting point $x_0 = 0.8$. Repeat for starting points $x_0 = 1$ and $x_0 = 1.1$. Does Newton's method always converge to a zero? Does it always converge to the nearest zero? Explain your answers.

3. Find a starting point for Newton's method that will cause it to get caught in a loop. That is $x_2 = x_0$. (Hint: If $x_{k+1} = N(x_k)$, then you want to solve $N(N(x)) = x$.)

4. Use the NEWTON.MTH file to make a picture of Newton's method with the starting point you found in Part 3.

Chapter 5

Riemann Sums and Integration

New *DERIVE* topics •Making sums •Calculating antiderivatives •Calculating definite integrals •Approximating definite integrals
Calculus concepts •Riemann sums •Antiderivatives •Definite integrals •Fundamental theorem of calculus

Conventions and Facts about Riemann Sums

To make Riemann sums for f on the interval $[a, b]$ using n subintervals, we will always use subintervals of equal length and will restrict our attention to left-hand sums (LHS) and right-hand sums (RHS) as defined below.

$$\text{Let } \Delta x = \frac{b-a}{n}. \text{LHS}_n = \sum_{i=0}^{n-1} f(a + i\Delta x)\Delta x \text{RHS}_n = \sum_{i=1}^{n} f(a + i\Delta x)\Delta x$$

1. If f is increasing on the interval $[a, b]$ then $\text{LHS}_n \leq \int_a^b f(x)\, dx \leq \text{RHS}_n$ (For decreasing functions, the inequality is reversed.)
2. For functions that are monotone (either increasing or decreasing) on $[a, b]$ the error in using either the left-hand sum or the right-hand sum to approximate the integral is no more than $|f(b) - f(a)|\Delta x$.

Solved Problem 5.1: The effect of n on Riemann sums

New *DERIVE* Lessons •Making sums •Calculating definite integrals

(a) Use right-hand sums with $n = 10$, 100, and 1000 to approximate $\int_1^3 \frac{1}{1+x^3}\, dx$.

(b) Use *DERIVE* to get the exact value of the integral and explain how the right-hand sums you calculated relate to this answer.

(c) Use the RIEMANN.MTH file in Appendix II to make a picture of the right-hand sums for $n = 5$ and $n = 10$. Give a geometric interpretation of these sums.

Solution to (a): First **Author** `F(x):=1/(1+x^3)`. For this integral we have $a = 1$ and $b = 3$ so that $\Delta x = \dfrac{3-1}{n} = \dfrac{2}{n}$. Thus the right-hand sum is $\sum_{i=1}^{n} f\left(1 + i\frac{2}{n}\right)\frac{2}{n}$. To make this, we **Author** `F(1+i(2/n))(2/n)` and use **Calculus Sum** setting the limits to Lower limit:1 Upper limit: n. (Press [C] for **Calculus**, and we will be presented with a new menu, one of whose items will be **Sum**. Press [S] and then Press [Enter] three times to accept the defaults. If it had been necessary to change the Upper limit: field, we would have used the [Tab] key.) *DERIVE* presents the right-hand sum in expression 3 of Figure 5.1.

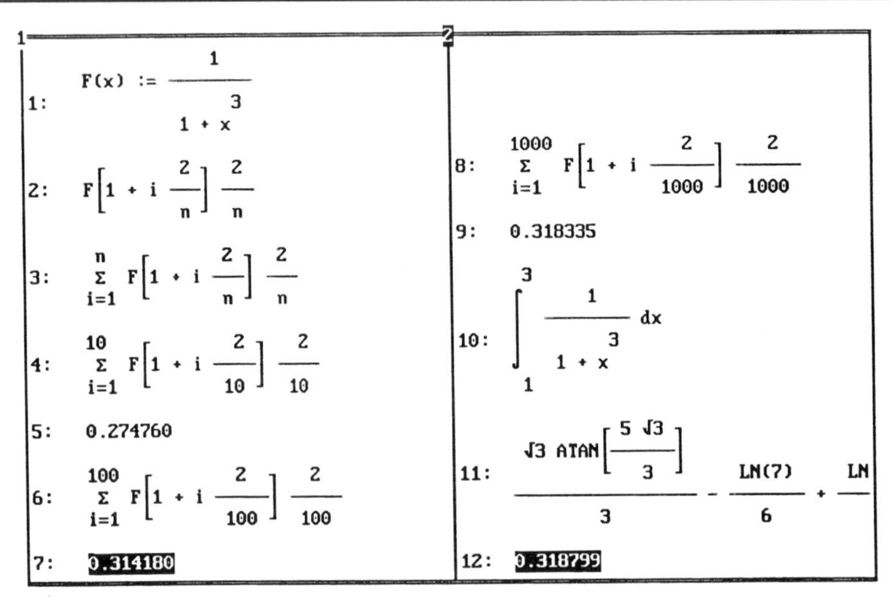

Figure 5.1: **Right-hand sums for** $\int_1^3 \frac{1}{1+x^3}\,dx$

We want to evaluate this sum at $n = 10$. With expression 3 highlighted, use **Manage Substitute**. Press [Enter] at the prompt MANAGE SUBSTITUTE value: i to indicate that i is not to be changed, but at the prompt MANAGE SUBSTITUTE value: n type `10` to replace n by 10. The result is in expression 4. **approX** to get the value of the sum in expression 5. To get the other two sums, highlight expression 3 of Figure 5.1 and use **Manage Substitute** to replace n by `100` and then by `1000`. The results are in expressions 7 and 9.

Solution to (b): Now we want to get the exact value of the integral. Use the arrow keys to highlight only the right-hand side of expression 1 in Figure 5.1. Use **Calculus Integrate** setting Lower limit:1 Upper limit:3. (Use [C] for calculus, and we see a new menu, one of whose items is **Integrate**. Press I. Press Enter twice to accept the defaults, and then set the limit fields.) The notation for the integral appears in expression 10. **Simplify** to get the answer in expression 11. This expression involves functions that we have not studied, so we **approX** to get the decimal value in expression 12. We note that as n increases, the estimates of the integral, 0.274760, 0.314180, 0.3318335, given by the right-hand sum increase toward the value of the integral, 0.318799. These are under-estimates as predicted by the decreasing graph in Figure 5.2.

Solution to (c): We assume you have saved the RIEMANN.MTH file according to the instructions in Appendix II. Load it now using **Transfer Merge** RIEMANN. This file has erased our definition of f, so we **Author** `F(x):=1/(1+x^3)` as seen in expression 22 of Figure 5.2. The

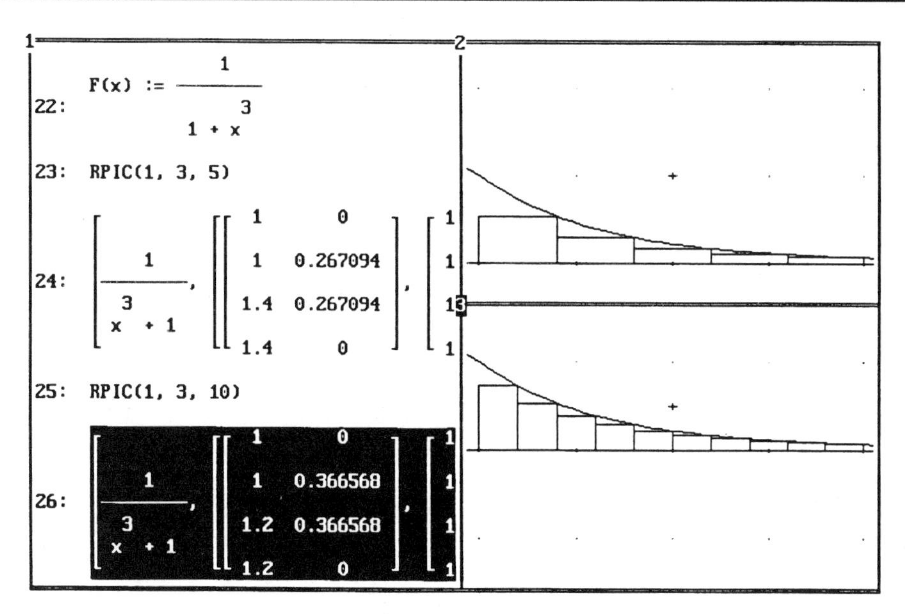

Figure 5.2: **A picture of right-hand sums, Scale x:0.5 y:0.5**

function RPIC(a, b, n) makes pictures of right-hand sums. (The LPIC(a, b, n) function makes pictures of left-hand sums.) We want a picture with $a = 1$, $b = 3$, and $n = 5$, so we **Author** `RPIC(1,3,5)` and then **approX**. The result in expression 24 tells *DERIVE* how to make a plot. **Plot** to open a plot window, but before making a picture we must change the default graphics settings using **Options State Rectangular Connected Small** Enter. (Do not press Enter until you have made all your selections.) Now **Plot**, and the picture of the right-hand sum appears in window 2 of Figure 5.2. We have zoomed in once to get a better picture. The integral is the area under the curve, and the right-hand sum is the total area of the boxes. The picture shows why the right-hand sum gives an under-estimate of the integral. Press A for **Algebra** to get back to the calculations. To make a picture using $n = 10$, **Author** `RPIC(1,3,10)`, **approX**, and **Plot**. The picture is in window 3 of Figure 5.2, and it shows why increasing n gives a better estimate of the area under the curve.

Solved Problem 5.2: Comparing left-hand sums and right-hand sums

New *DERIVE* Lessons •Approximating definite integrals

Let $f(x) = \sin(x^2)$.

(a) Plot the graph of f to verify that it is increasing on the interval $[0, 1]$.

(b) Calculate the left-hand sums for $\int_0^1 \sin(x^2)\, dx$ with $n = 10$ and $n = 100$.

(c) Calculate the right-hand sums for $\int_0^1 \sin(x^2)\, dx$ with $n = 10$ and $n = 100$.

(d) Calculate the integral and explain how your estimates in (b) and (c) relate to this value.

(e) Use the RIEMANN.MTH file to make pictures of the left-hand sum and right-hand sum with $n = 10$ and explain how they relate to the integral.

Solution to (a): **Author** `F(x):=SIN(x^2)` and **Plot**. The graph is in window 2 of Figure 5.3, and it is clearly increasing on the interval $[0, 1]$.

Solution to (b): We have $a = 0$ and $b = 1$ so that $\Delta x = \frac{1}{n}$. The left-hand sum is $\sum_{i=0}^{n-1} f\left(i\frac{1}{n}\right)\frac{1}{n}$. To make it, **Author** `F(i/n)(1/n)` and use **Calculus Sum** setting Lower limit:`0` Upper limit:`n-1`. The sum appears in expression 3 of Figure 5.3. To get the left-hand sum with 10 subintervals, highlight expression 3 and use **Manage Substitute** to replace n by `10`. **approX** to get the result in expression 5. Repeat this procedure replacing n by `100` to get the left-hand sum for 100 subintervals in expression 7.

Solution to (c): The right-hand sum is $\sum_{i=1}^{n} f\left(i\frac{1}{n}\right)\frac{1}{n}$. Highlight expression 2 of Figure 5.3 and use **Calculus Sum** setting Lower limit: `0` Upper limit: `n`. The right-hand sum appears in expression 8 of Figure 5.4. We get the right-hand sums for $n = 10$ and $n = 100$ subintervals by using **Manage Substitute** and replacing n by these values. The results are in expressions 10 and 12.

Solution to (d): To get the integral, highlight expression 1 and use **Calculus Integrate** setting Lower limit:`0` Upper limit:`1`. The integral appears in expression 13 of Figure 5.4. When

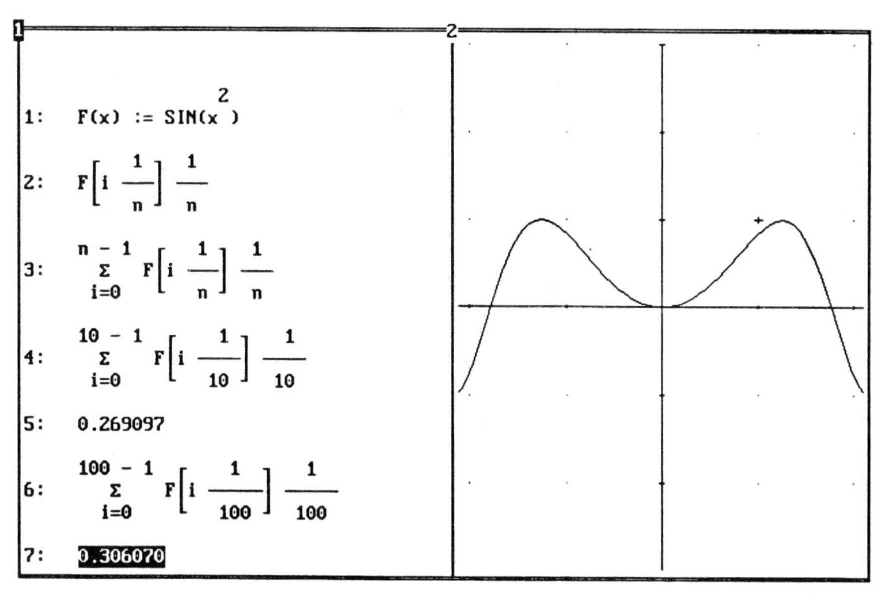

Figure 5.3: **Left-hand sums for** $\int_0^1 \sin(x^2)\,dx$

we ask *DERIVE* to **Simplify**, it returns the integral unevaluated as seen in expression 14. This means that *DERIVE* is not able to calculate the exact value of this integral, but it can provide an approximation. We **approX** expression 14, to get the answer correct to six digits in expression 15.

Since our function is increasing on the interval $[0, 1]$ we expect left-hand sums to be too small and right-hand sums to be too large. Furthermore, we expect both sums to give more accurate answers when n is large. This is verified by the fact that

$$0.269097 < 0.306070 < 0.310267 < 0.314484 < 0.353244$$

Solution to (e): To make the pictures, we load the RIEMANN.MTH file using **Transfer Merge** RIEMANN. **Author** `F(x):=SIN(x^2)`, and then make the picture of the left-hand sum with 10 subintervals using `LPIC(0,1,10)`. To see the picture in window 2 of Figure 5.5, **approX** and **Plot**. (Do not forget to reset the default graphics options using **Options State Rectangular Connected Small**.) We make the right-hand sum with `RPIC(0,1,10)`. Its picture appears in window 3. Since the integral is the area under the curve, these pictures make clear why the left-hand sum gives an under-estimate and the right-hand sum gives an over-estimate.

```
8:     n     [    1 ] 1
       Σ   F [ i ── ] ──
      i=1    [    n ] n

       10    [    1  ]  1
9:     Σ   F [ i ──  ] ──
      i=1    [   10  ] 10

10:  0.353244

      100    [    1   ]  1
11:    Σ   F [ i ───  ] ───
      i=1    [   100  ] 100

12:  0.314484
```

```
       1
       ⌠                   2
13:    ⌡  (F(x) := SIN(x  )) dx
       0

       1
       ⌠        2
14:    ⌡  SIN(x  ) dx
       0

15:  0.310267
```

Figure 5.4: **Right-hand sums for** $\int_0^1 \sin(x^2)\, dx$

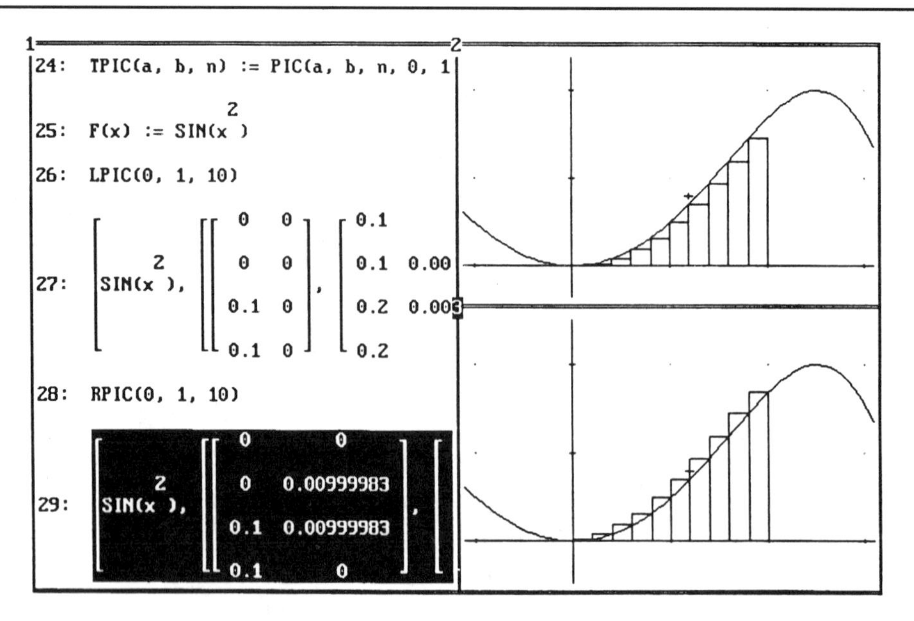

Figure 5.5: **Pictures of left-hand sums and right-hand sums for** $\int_0^1 \sin(x^2)\, dx$

Solved Problem 5.3: Getting approximations to desired degrees of accuracy

How large must you choose n so that the left-hand or right-hand Riemann sum approximates $\int_1^4 \cos\left(\frac{1}{x}\right)\,dx$ with error less than 0.001? Obtain the approximation and compare it with *DERIVE's* calculation of the integral.

Solution: First **Author** `F(x):=COS(1/x)` and **Plot** to get the graph in window 2 of Figure 5.6. Since f is increasing on the interval $[1,4]$, the error in approximating the integral with either the left-hand sum or the right-hand sum is no more than $|f(b)-f(a)|\Delta x$. Thus we want to choose n so that $|f(b)-f(a)|\Delta x \leq 0.001$. We have $a=1$, $b=4$, and $\Delta x = \frac{3}{n}$. We **Author** `|F(4)-F(1)|(3/n)<0.001`. **approX** and then **soLve**. The result we want is in expression 5 of Figure 5.6. When we **approX** this, we see that n should be at least 1286.

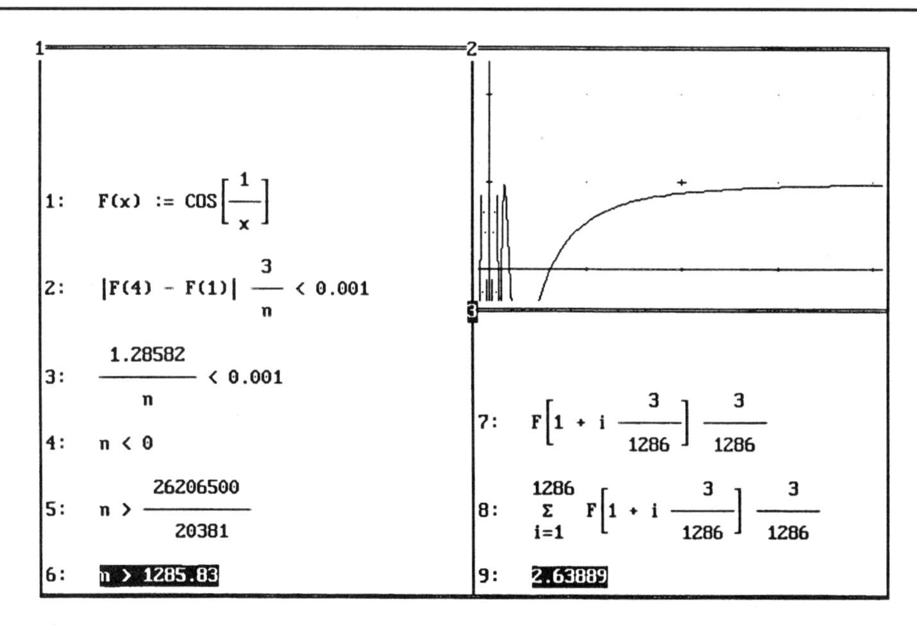

Figure 5.6: **Choosing n to make the error less than** 0.001

We will use a right-hand sum, and since our function is increasing, we know that this will give an over-estimate of the integral. The sum, as seen in expression 8 of Figure 5.6, is $\sum_{n=1}^{1286} F\left(1+i\frac{3}{1286}\right)\frac{3}{1286}$. We **approX** to get its value in expression 9. To get the integral in

```
1
10:  ∫ [F(x) := COS[1/x]] dx   (from 1 to 4)

11:  2.63839

12:  |2.63839 - 2.63889|

13:  4.98796 10^-4
```

Figure 5.7: **The error in approximating** $\int_1^4 \cos\left(\frac{1}{x}\right)\, dx$ **with a right-hand sum**

expression 10 of Figure 5.7, we highlight expression 1 and use **Calculus Integrate** setting Lower limit:1 Upper limit:4. This is an integral that *DERIVE* cannot evaluate exactly, so we **approX** to get expression 11. In expression 13 we have calculated the actual error, 0.000498796, in our estimate, and it is less than 0.001 as required.

Practice Problems

1. Calculate $\int_0^{\pi} \sin x\, dx$. Answer: 2

2. Approximate $\int_0^1 \frac{1}{2^x + 3^x}\, dx$. Answer: 0.328502

3. Evaluate the left-hand sum for $f(x) = \frac{1}{1+x}$ on $[0,\ 2]$ with 78 subintervals. Answer: 1.10720

4. If $f(x) = \frac{2}{3+x^2}$, $a = 3$, $b = 5$, and $\Delta x = \frac{b-a}{n}$, solve the inequality $|f(b) - f(a)|\Delta x < 0.01$ for the positive integer n. Answer: $n > 20$

5. Calculate $\int \cos\sqrt{x}\, dx$ Answer: $2\cos\sqrt{x} + 2\sqrt{x}\sin\sqrt{x} + c$

Laboratory Exercise 5.1

Riemann Sums and the Number of Subintervals

Name ______________________________ **Due Date** __________

Consider $\int_1^5 \frac{1+x^2}{1+x}\,dx$.

1. Calculate the right-hand sums for the integral using $n = 10$, $n = 100$, and $n = 1000$ subintervals.

2. Calculate the left-hand sums for the integral using $n = 10$, $n = 100$, and $n = 1000$ subintervals.

3. Plot the graph of $\frac{1+x^2}{1+x}$ and determine if it is increasing or decreasing on the interval $[1,5]$. Based on this information and on your answers in Parts 1 and 2, give your best lower and upper estimates for the integral. Explain how you got your answers.

4. Use *DERIVE* to calculate the exact value of the integral and verify your answer in Part 3.

5. Use the RIEMANN.MTH file in Appendix II to make pictures of the left-hand sum and right-hand sum for $n = 10$.

6. Give a geometric interpretation of the pictures in Part 5, and explain how they relate to the integral.

Laboratory Exercise 5.2

Riemann Sums and Monotonicity

Name ______________________________ **Due Date** ____________

1. Calculate the left-hand sum for $\int_0^\pi \sin x\, dx$ using 10 subintervals and explain how your answer relates to the exact value of the integral.

2. Use the RIEMANN.MTH file to make a picture of the sum from Part 1 and use the picture to give a geometric explanation of the relationship you found there.

3. Calculate the left-hand sum for $\int_{\pi}^{2\pi} \sin x \, dx$ using 10 subintervals and explain how your answer relates to the exact value of the integral.

4. Use the RIEMANN.MTH file to make a picture of the sum from Part 3 and use the picture to give a geometric explanation of the relationship you found there.

Laboratory Exercise 5.3

Controlling the Error in Riemann Sums

Name ______________________________ **Due Date** ____________

Let $f(x) = \dfrac{x}{1+x^3}$.

1. Verify that f is monotone on the interval $[1, 3]$. Is it increasing or decreasing?

2. How many subintervals are needed to ensure that a right-hand sum approximates $\int_1^3 f(x)\,dx$ with error no more than 0.001? Explain how you got your answer.

3. Calculate the right-hand sum using the value of n you found in Part 2. Use *DERIVE* to evaluate the integral and verify that the error is no more than 0.001.

4. Find a number c between 0 and 3 with the property that $f(x)$ is increasing on $[0, c]$ and decreasing on $[c, 3]$. Explain how you got your answer.

5. Use Riemann sums to approximate $\int_0^3 f(x)\, dx$ with error no more than 0.01. Explain how you got your answer and why you are certain of your error bound. (Hint: The number c you found in Part 4 should be helpful. Remember that if two approximations are added, the error in the sum may be as large as the sum of the errors for each summand.)

6. Use *DERIVE* to evaluate $\int_0^3 \frac{x}{1+x^3}\, dx$ and verify that your answer in Part 5 is within 0.01 of the correct answer.

Laboratory Exercise 5.4

Integrals as Limits of Riemann Sums

Name ______________________________ **Due Date** __________

Let $f(x) = x^4 - x^3$.

1. Make a right-hand sum for $\int_0^2 f(x)\,dx$ using n subintervals and use *DERIVE* to **Simplify** it.

2. Calculate the limit as n goes to infinity of your answer in Part 1.

3. Make a left-hand sum for $\int_0^2 f(x)\,dx$ using n subintervals and use *DERIVE* to **Simplify** it. Did you get the same answer as in Part 1?

4. Calculate the limit as n goes to infinity of your answer in Part 3.

5. Are your answers in Parts 2 and 4 the same? Explain how they relate to $\int_0^2 f(x)\,dx$.

Chapter 6
Applications of the Integral

New *DERIVE* topics •Defining constants
Calculus concepts •Areas •Solids of revolution •Arc length •Surface area •Rectilinear motion •Work •Center of mass

Solved Problem 6.1: Integrals as areas

Plot the graphs of $\sin x$ and x^4 on the same screen and calculate the area enclosed between the two curves.

Solution: **Author** `SINx` and `x^4` and **Plot** them as seen in window 2 of Figure 6.1. The area we are looking for is the crescent shaped region between the two graphs in the first quadrant. The graphs cross at the origin and once more near $x = 1$. To find this point we solve $\sin x = x^4$ or, equivalently, $\sin x - x^4 = 0$. The latter will be most convenient for the work that follows. **Author** `#1-#2`, change to the approximate mode using **Options Precision Approximate**, and **soLve** on the range Lower:`0.5` Upper:`1`. The solution is in expression 4. Since the graph of the sine function is on top, we need to calculate $\int_0^{0.949616} \sin x - x^4\, dx$.

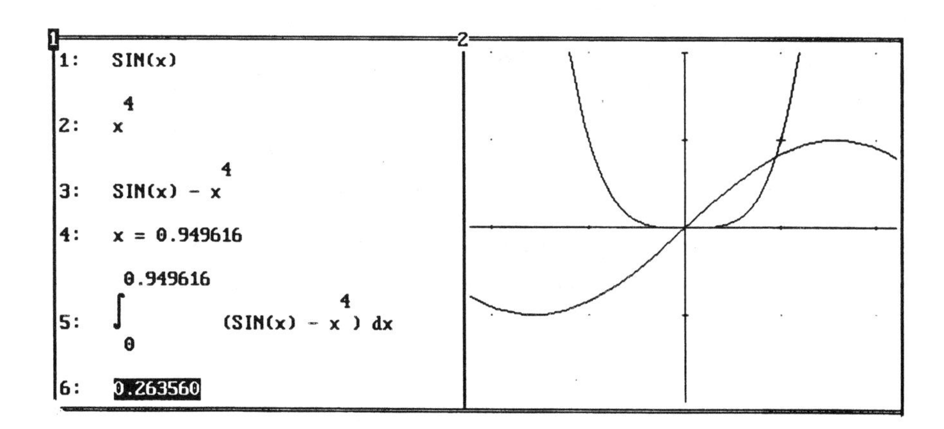

Figure 6.1: **The area enclosed by** $\sin x$ **and** x^4

Highlight expression 3 of Figure 6.1 and use **Calculus Integrate** setting Lower limit:`0` Upper limit:`0.949616`. The integral appears in expression 5. Since we are still in the approximate mode, either **Simplify** or **approX** will produce the answer in expression 6.

Solved Problem 6.2: Volume of solids of revolution

Find the volume of the solid generated by revolving the region in the first quadrant enclosed by $2\sin x - x$ and the x-axis around the x-axis. Find the volume if the region is revolved around the y-axis.

Solution: We first **Author** `2SINx-x` and **Plot**. From the graph in window 3 of Figure 6.2, the region we want to use is the "bump" in the first quadrant. We will use the "disk method" to get

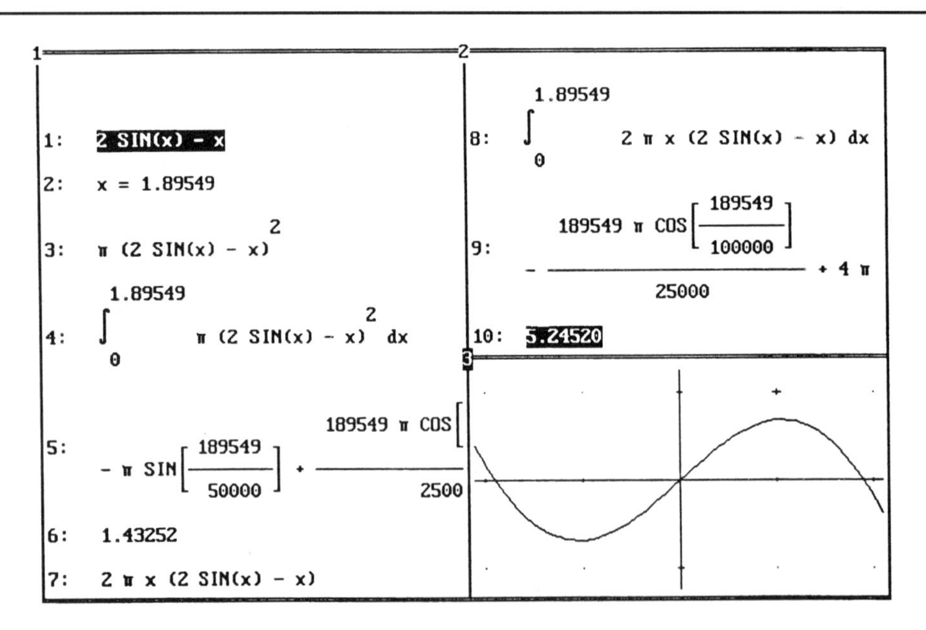

Figure 6.2: **Volume of a solid of revolution**

the volume of the region revolved around the x-axis. Thus we need to integrate $\pi(2\sin -x)^2$ from $x = 0$ to the point where the graph crosses the positive x-axis. This clearly occurs between $x = 1$ and $x = 2$, so we change to the approximate mode using **Options Precision Approximate** and **soLve** expression 1 on the range Lower:1 Upper:2. The solution appears in expression 2. To get the integral we **Author** `pi #1^2` and then use **Calculus Integrate** setting Lower limit:`0` Upper limit:`1.89549`. The integral appears in expression 4. We change back to the exact mode using **Options Precision Exact** to evaluate it. **Simplify** to get the exact answer partially displayed in expression 5, and then **approX** to get a decimal approximation. (Note: In this case, the change back to exact mode is of questionable value. If we had remained in the approximate mode, *DERIVE* would have skipped expression 5 and gone directly to a decimal approximation of the integral, but a small amount of accuracy would be lost.)

We use the "shell method" to get the volume of the region revolved around the y-axis. Thus we wish to integrate $2\pi x(2\sin x - x)$. **Author** `2pi x#1` and use **Calculus Integrate** setting Lower limit:`0` Upper limit:`1.89549` to get the integral in expression 8 of Figure 6.2. Its exact value is partially displayed in expression 9, and its decimal approximation is in expression 10.

Solved Problem 6.3: Arc length

(a) Find the length of the arc of the graph of $\sin x$ from $x = 0$ to $x = \pi$.

(b) Find a so that the length of the arc of the graph of $\sin x$ from $x = 0$ to $x = a$ is 1 unit.

Solution to (a): First **Author** `SINx` and then `SQRT(1+DIF(#1,x))`. **Simplify** to get the integrand in expression 3 of Figure 6.3. Now use **Calculus Integrate** setting the limits to Lower

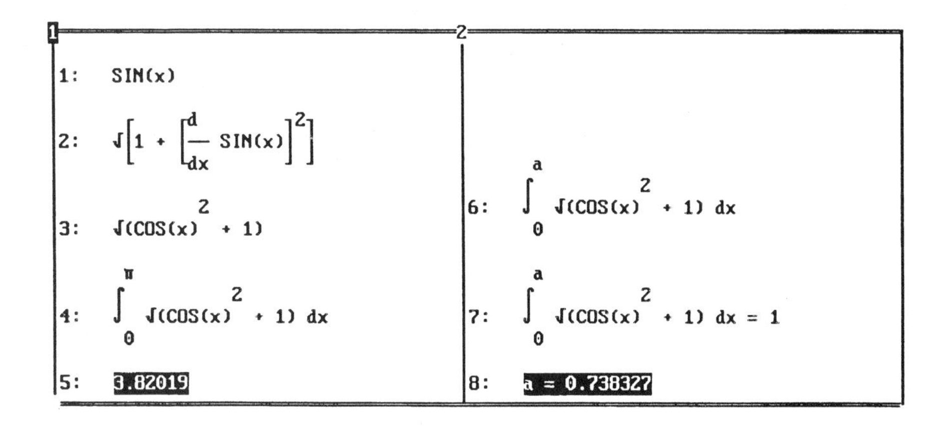

Figure 6.3: **Calculating arc length**

limit: `0` Upper limit: `pi`. The integral that gives arc length is in expression 4. *DERIVE* cannot evaluate this integral exactly, and so we **approX** to get the arc length in expression 5.

Solution to (b): The arc length from $x = 0$ to $x = a$ is given by the integral from 0 to a of expression 2 which appears in expression 6 of Figure 6.3. We want this length to be 1, so we **Author** `#6=1`. We know that $a = 0$ gives an arc length of 0, and $a = \pi$ gives an arc length of 3.82019. Thus the value of a we are looking for is somewhere between 0 and π. We change to the approximate mode using **Options Precision Approximate** and **soLve** equation 8 on the range Lower: `0` Upper: `pi`. The solution is in expression 8.

Solved Problem 6.4: Rectilinear motion

An object moves along the x-axis so that its acceleration at time t is given by $a(t) = t \sin t$. At time $t = 1$ the point is located at $s(1) = 2$, and its velocity is $v(1) = 3$.

(a) Find the velocity function, $v(t)$, and the position function, $s(t)$.

(b) Find the total distance traveled by the object from time $t = 0$ to time $t = 5$.

Solution to (a): The velocity function $v(t)$ is an antiderivative of the acceleration function. Thus we **Author** `tSINt` and use **Calculus Integrate** to get the antiderivative in expression 3 of Figure 6.4. (We have split the window there in an unusual fashion to conserve space.) The role of the constant is crucial here; $v(t) = \sin t - t \cos t + c$, and we need to evaluate c. We know that $v(1) = 3$, so that $\sin 1 - \cos 1 + c = 3$. Or, solving for c, $c = 3 - \sin 1 + \cos 1$. The result is in expression 5, and we **Author** `#3+#5` to get the velocity function in expression 6.

We integrate expression 6 to get the position function, $s(t)$. The result, without the necessary constant, is in expression 9 of Figure 6.4. We know that $s(1) = 2$, so $-2\cos 1 - \sin 1 + 2.69883 + c = 2$, and thus $c = 2 + 2\cos 1 + \sin 1 - 2.69883$. This is calculated in expressions 10 and 11; and $s(t)$ appears in expression 12.

Solution to (b): In general, the distance traveled is given by the integral of the absolute value of the velocity function. (The absolute value of velocity is speed.) Thus we **Author** `|#12|` and then use **Calculus Integrate** setting Lower limit:`0` Upper limit:`5`. We **approX** the integral to get the distance traveled in expression 15 of Figure 6.4.

```
1:   t SIN(t)

2:   ∫ t SIN(t) dt

3:   SIN(t) - t COS(t)

4:   3 - SIN(1) + COS(1)

5:   2.69883

6:   SIN(t) - t COS(t) + 2.69883

7:   ∫ (SIN(t) - t COS(t) + 2.69883) dt

                                   7160 t
8:   - 2 COS(t) - t SIN(t) + ──────
                                    2653

9:   - 2 COS(t) - t SIN(t) + 2.69883 t

10:  2 + 2 COS(1) + SIN(1) - 2.69883

11:  1.22324

12:  - 2 COS(t) - t SIN(t) + 2.69883 t + 1.22324

13:  |- 2 COS(t) - t SIN(t) + 2.69883 t|

      5
14:  ∫  |- 2 COS(t) - t SIN(t) + 2.69883 t| dt
      0

15:  39.5054
```

Figure 6.4: **Recovering position from acceleration**

Solved Problem 6.5: From the Earth to the Moon

New *DERIVE* Lessons •Defining constants

(a) Find the work required to move a rocket ship of mass m from the surface of the Earth to the surface of the Moon.

(b) What initial velocity must be given to a cannonball fired from the surface of the Earth if it is to land on the Moon?

Solution to (a): In our solution we are going to take into account the influence of the Moon's gravity. The following data will be necessary.

$G =$	Universal gravitational constant	$= 6.67 \times 10^{-11}$
$E =$	Mass of the Earth	$= 5.98 \times 10^{24}$ kilograms
$P =$	Radius of the Earth	$= 6.38 \times 10^{6}$ meters
$L =$	Mass of the Moon	$= 7.35 \times 10^{22}$ kilograms

$Q =$	Radius of the Moon	$= 1.74 \times 10^6$ meters
$D =$	Distance from center of Earth to center of Moon	$= 3.84 \times 10^8$ meters

We can define constants just as we do functions using `:=`. Thus if we **Author** `g:=6.67 10^(-11)`, *DERIVE* will know that g stands for the number 6.67×10^{-11}. In expression 1 of Figure 6.5, we have defined all the constants in the table above using a list enclosed by square brackets and separated by commas to save space. (You may find it easier to define each constant on a separate line or to leave them undefined and use **Manage Substitute** at the end of your work.) We also changed to the approximate mode using **Options Precision Approximate**.

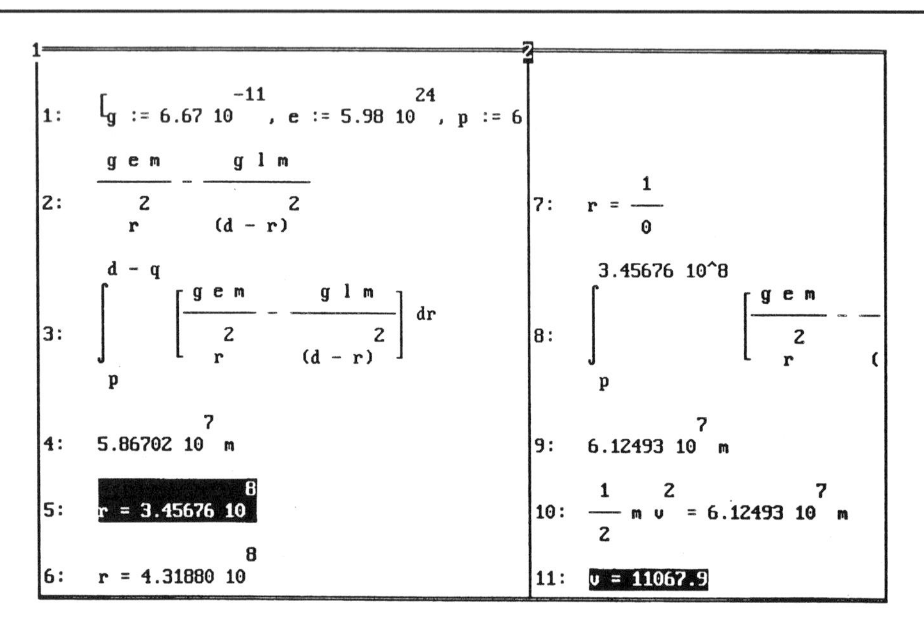

Figure 6.5: **Work required to get from the Earth to the Moon**

If r is the distance from the center of the Earth to the rocket ship, then the force pulling it toward the Earth is $\dfrac{GEm}{r^2}$ while the force pulling it toward the Moon is given by $\dfrac{GLm}{(D-r)^2}$. Thus the net force on the rocket is the difference $\dfrac{GEm}{r^2} - \dfrac{GLm}{(D-r)^2}$.

To obtain the work, we integrate the force from the surface of the Earth, $r = P$, to the surface of the Moon, $r = D - Q$. (When you use **Calculus Integrate**, be sure to set the variable: field to `r`.) The resulting integral appears as expression 3 of Figure 6.5. We **approX** to get the work, $5.86702 \times 10^7 m$ joules, in expression 4. (Recall that a joule is a Newton-meter of work.)

Solution to (b): The cannonball of mass m must have sufficient kinetic energy, $\frac{1}{2}mv^2$, to pass the point where the gravitational attraction of the Moon and the Earth are equal. From that point it will fall to the surface of the Moon. To find that point, we highlight expression 2 and **soLve** setting the Solve variable: to `r`. We see three solutions in expressions 5, 6, and 7 of Figure 6.5. Expression 7 is $\frac{1}{0}$, which is nonsense, and expression 6 is greater than the distance from the Earth to the Moon, so it cannot be correct. We conclude that the equilibrium point occurs at $r = 3.45676 \times 10^8$ meters. The work required to lift the cannonball to this point is

$$\text{Work} = \int_p^{3.45676\times 10^8} \frac{GEm}{r^2} - \frac{GLm}{(D-r)^2}\, dr$$

Use **Calculus Integrate** once more to integrate expression 2 from `P` to `3.45676 10^8` setting the Integrate variable: field to `r`. **Simplify** to obtain the work, $6.12493 \times 10^7 m$ joules, as seen in expression 9.

We get the required velocity by setting the kinetic energy equal to this work and solving. **Author** `1/2 mv^2=#9` and **soLve**, setting the Solve variable to `v`. The result is 11,067.9 meters per second seen in expression 11.

Solved Problem 6.6: Area and center of mass

A thin sheet of metal is shaped like the region in the first quadrant between $y = \sin x$ and $y = \frac{x}{2}$.

(a) Find the area of the region.

(b) Find the x-coordinate of the center of mass of the region.

Solution to (a): We **Author** and **Plot** both `SIN x` and `x/2` as seen in window 3 of Figure 6.6. From the graph we see that to find the area we will have to integrate $\sin x - \frac{x}{2}$ from $x = 0$ to the point where the graphs cross to the right of $x = 0$. Therefore, we must solve $\sin x = \frac{x}{2}$ or, equivalently, $\sin x - \frac{x}{2} = 0$. This cannot be done exactly, so we use **Options Precision Approximate** and then **soLve** on the range Lower: `1` Upper: `3`. The solution is seen in expression 4. Thus the area is

$$\text{Area} = \int_0^{1.89549} \sin x - \frac{x}{2}\, dx$$

Highlight expression 3 of Figure 6.6 and use **Calculus Integrate** setting Lower limit: `0` Upper limit: `1.89549`. When we **approX**, we get the area, 0.420798, as expression 6 of Figure 6.6.

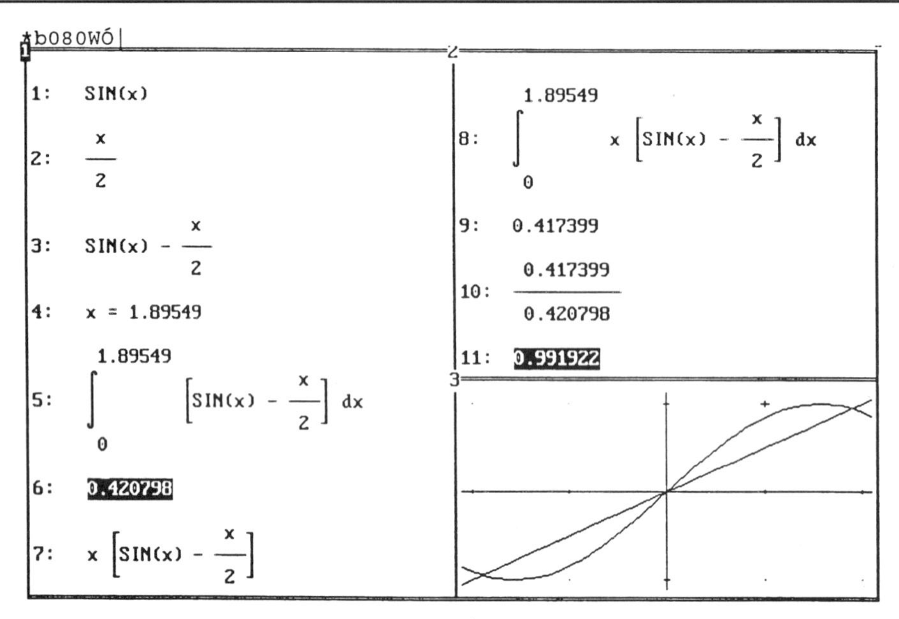

Figure 6.6: **Area and center of mass**

Solution to (b): The x-coordinate of the center of mass of the region bounded above by $y = f(x)$ and below by $y = g(x)$ on the interval $[a, b]$ is given by the formula

$$\frac{\int_a^b x(f(x) - g(x))\,dx}{\int_a^b f(x) - g(x)\,dx}$$

The numerator is called the "first moment" of the region about the y-axis. The denominator is the area. If we placed this region on a thin rod parallel to the y-axis at this x-value, the region would balance on the rod. To finish this problem, we need to find $\int_0^{1.89549} x\left(\sin x - \frac{x}{2}\right)\,dx$ and divide it by the area we found in (a). Thus, we **Author** `x #3` and use **Calculus Integrate** setting Lower limit: `0` Upper limit: `1.89549` to get expression 8. When we **approX**, we get the moment, 0.417399, as expression 9 in Figure 6.6. The ratio, 0.991922, appears as expression 11. If the line $x = 0.991923$ were a thin rod, the region would balance on the rod.

Practice Problems

1. Find the area that is enclosed by $\sin x$ and the x-axis from $x = 0$ to $x = \pi$. Answer: 2

2. Find the x-coordinate of the point of intersection of the graphs of $\dfrac{1}{1+x^2}$ and x. Answer: 0.682327

Laboratory Exercise 6.1

Calculating Areas

Name ______________________________ Due Date ____________

Produce pictures of the following regions and calculate their areas. Your answers should include the appropriate integral, its exact value where possible, and an approximation.

1. The region enclosed by the graphs of $\cos x$ and x^2.

2. The region enclosed by the graphs of $\sqrt{1+x^4}$ and $2\cos x$.

3. The region inside the unit circle and above the graph of x^5. (Note: If you look carefully at the picture, you can get the exact answer without calculus or a computer.)

4. All the regions enclosed by the graphs of $\sin(6x)$ and $1 - x^2$. (Hint: Recall that $\int_a^b |f(x) - g(x)|\, dx$ gives the area between the graphs of $f(x)$ and $g(x)$ from $x = a$ to $x = b$ no matter which graph is on top.)

Laboratory Exercise 6.2

Solids of Revolution and Surface Area

Name ______________________________ Due Date ____________

For Parts 1 through 4, find the volume of the solid obtained by revolving the given regions about the x-axis and then about the y-axis. State the method you use, show the integral, and give its value. Find the surface area if the region is revolved around the x-axis.

1. The "triangular" region enclosed by the graph of $\cos x - \sqrt{x}$, the x-axis, and the y-axis.

2. The region enclosed by the graphs of $\cos x$ and $(x-1)^2$.

3. The region in the first quadrant enclosed by the graphs of $\csc x$ and $\dfrac{2\sin x}{x}$.

4. The region enclosed by the graphs of $\cos x$ and $x^2 - 2x$. (Warning: Be careful of overlap! Look carefully at the picture and decide what you want to revolve.)

5. Find the volume and surface area obtained by rotating the upper half of the unit circle around the line $x = 3$.

6. Find the volume of a doughnut whose inner radius is r and whose outer radius is R. Find the surface area in case $r = 3$ and $R = 5$. (Hint: First express the doughnut as a solid of revolution.)

Laboratory Exercise 6.3

Applications of Arc Length

Name ______________________________ Due Date ___________

1. The equation of an ellipse is $\frac{x^2}{a^2} + \frac{y^2}{b^2} = 1$.

   (a) Show that the area of the ellipse is given by πab.

   (b) Write the integral that gives the circumference of an ellipse and ask *DERIVE* to evaluate it. What happens and what do you conclude?

   (c) Find the circumference of the ellipse when $a = 2$ and $b = 3$.

2. A cannonball is fired from the origin and follows the graph of $x(0.8 - 0.01x)$. Assuming the x-axis represents ground level, how far did the cannonball travel?

3. A ship starts at the point $(0, 30)$ and navigates into the first quadrant along the graph of $y = 30\sqrt{1 + \frac{x^2}{600}}$. The ship only has enough fuel to go 500 miles. Where will it be when it runs out of fuel?

Laboratory Exercise 6.4

The Efficiency of a Fence

Name ______________________________ **Due Date** ____________

A length of fence stretches above the x-axis with its ends at $(-1, 0)$ and at $(1, 0)$. We can measure the "efficiency" of the fence by

$$\text{Efficiency} = \frac{\text{Area}}{\text{Length}}$$

where "Area" is the area enclosed by the fence above the x-axis. The larger this ratio, the more area we are enclosing per unit length of fence. Plot the graphs of the following fences and find the efficiency of each.

1. $1 - x^2$

2. $(1 - x^2)^2$

3. $\cos\left(\frac{\pi x}{2}\right)$

4. Design a fence that is more efficient than any of the three fences above. (If you can think of a strategy for doing it, explain what it is.)

Laboratory Exercise 6.5

Applications of Rectilinear Motion

Name ______________________________ Due Date ____________

1. The downward velocity of a parachutist t seconds after jumping from an airplane is given by $v(t) = 20(1 - (0.2)^t)$ feet per second. If the parachutist lands on the ground 58 seconds after jumping, what was the altitude of the airplane? Show your work.

2. A spring vibrates so that the velocity of its tip is given by $v(t) = (0.8)^t(9\cos(5t) - 15\sin(5t))$ centimeters per second. Find the total distance the spring traveled in the first 8 seconds. Show your work.

3. Locate the Earth at the origin, and an asteroid far down the positive x-axis. The asteroid is falling toward Earth so that its velocity t hours after it was first observed is given by

$$v(t) = -\frac{186000}{(8760 - t)^{\frac{1}{3}}} \text{ kilometers per hour}$$

(a) How far did the asteroid travel during the first six months?

(b) Plot the graph of $v(t)$ and determine when the asteroid will strike Earth. Also discuss the validity of the formula for $v(t)$ when the asteroid is very near the Earth.

(c) Find a formula for the position function $s(t)$. (Hint: To evaluate the constant, you need to take into account that the asteroid is at the origin at the time you found in (b).)

(d) How far away was the asteroid when it was first observed?

Laboratory Exercise 6.6

From the Earth to the Sun

Name ______________________ Due Date ____________

1. Taking into account the gravitational pull of the Sun, how much work is required to send a rocket ship of mass m from the surface of the Earth to the surface of the Sun? You will need some of the data in Solved Problem 6.5 as well as the following data about the Sun.

$$\begin{aligned} \text{Mass of Sun} &= 1.97 \times 10^{30} \text{ kilograms} \\ \text{Radius of Sun} &= 6.95 \times 10^{8} \text{ meters} \\ \text{Distance from Earth to Sun} &= 1.49 \times 10^{11} \text{ meters} \end{aligned}$$

2. Explain what the sign of the answer means.

3. Find the point between the Earth and the Sun where the pull of the Earth's gravity matches that of the Sun.

4. Calculate the initial velocity of a cannonball fired from the surface of the Earth if it is to land on the surface of the Sun.

5. Calculate the escape velocity for the Sun. That is, what initial velocity must be given to a cannonball fired from the surface of the Sun so that it will never fall back to the Sun?

Laboratory Exercise 6.7

The Center of Mass of a Sculpture

Name ______________________ **Due Date** ____________

An art student was given a circular metal disk 2 feet in diameter and told to drill a small hole in it so that when the disk was cut in half and the piece with the hole was placed atop a spike stuck in the hole, it would balance. Not knowing about integrals, the artist drilled a hole at a point halfway between the center and the edge.

1. Where *should* the art student have drilled the hole? (Hint: Consider the right half of the circle $x^2 + y^2 = 1$, and see Solved Problem 6.6 for a discussion of computing a center of mass.)

2. Now that the student has made the mistake, he decides that rather than drill a second hole, he will cut the piece with the hole in it in such a way that it will balance on the spike at the point of the hole. Explain clearly how the disk should be cut so that our artist friend can understand.

3. What is the area of the piece of the metal disk that balances at the point where the hole $\left(\frac{1}{2}, 0\right)$ was drilled?

Chapter 7

Logarithmic and Exponential Functions

New *DERIVE* topics •Graphing inverses of functions
Calculus concepts •Inverse functions •Logarithms •Exponential functions

Solved Problem 7.1: Inverses of functions

New *DERIVE* Lessons •Graphing inverses of functions

Decide whether the functions (a) $f(x) = x^5 - x + 1$ and (b) $f(x) = x^5 + x + 1$ have inverses. If the inverse exists, find $f^{-1}(3)$ and $f^{-1}(2)$, and plot the graph of the inverse function.

Solution to (a): **Author** `x^5-x+1` and **Plot** to obtain the graph in window 2 of Figure 7.1. Graphically we are looking to see if f passes the *horizontal line test.* That is, we are checking to see if each horizontal line crosses the graph of f at most once. We have included the graph of the horizontal line $y = 1$ in window 2, which crosses the graph of f three times. Thus f has no inverse.

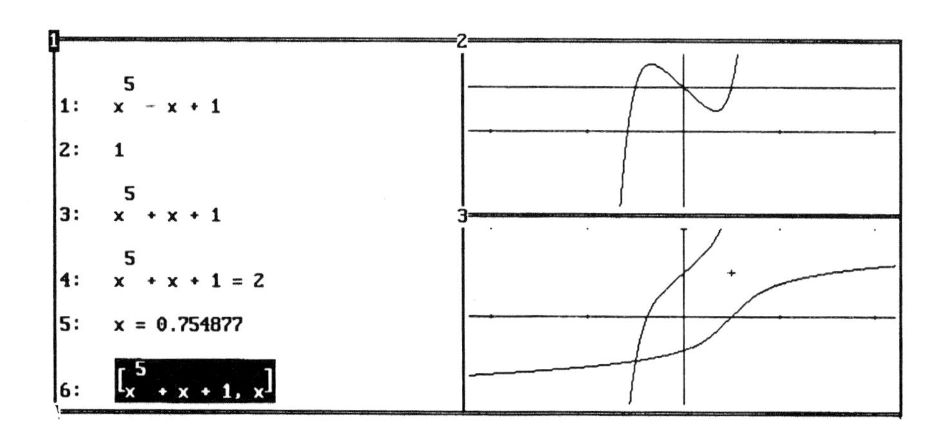

Figure 7.1: **The graph of the inverse of f. Scale x:2 y:2**

Solution to (b): The graph of $x^5 + x + 1$ appears in window 3 of Figure 7.1. (Which of the two is it?) It appears that the graph of f is increasing and thus passes the horizontal line test, indicating that the inverse exists. (You should be able to verify that f is increasing using the first derivative.)

To find $f^{-1}(3)$, we need to find x so that $x^5 + x + 1 = 3$. We don't need a computer to do that; $x = 1$ clearly works. Thus $f^{-1}(3) = 1$. But to find $f^{-1}(2)$, we need to solve the harder equation $x^5 + x + 1 = 2$. **Author** `#3=2`. From the graph in window 2, we see that $f(x) = 2$ near $x = 1$. Thus we change to the approximate mode using **Options Precision Approximate** and **soLve** on the range Lower:`0` Upper:`2`. From expression 5, we conclude that $f^{-1}(2) = 0.754877$.

Even though we do not have an explicit formula for f^{-1}, we can still plot its graph. The graph of a function f with domain D is the set of ordered pairs $(x, f(x))$ with x in D. If f has an inverse on D, then the graph of its inverse is the set of ordered pairs $(f(x), x)$, over a suitably restricted interval. Therefore, to get the graph of f^{-1}, we **Author** `[#3, x]`, and **Plot**. We see PLOT: Min:−3.1416 Max 3.1416, which is asking for an interval on which to plot the ordered pairs. We want to display the graph across as much of the width of the screen as we can, so we choose Min: `-2` Max: `2`. The graph of f^{-1} is the lower graph in window 3 of Figure 7.1.

DERIVE's Syntax for Logarithms and the Number e

To enter the special number e, you must **Author** [Alt e]. It will appear as $\hat{e}$. If you just **Author** `e`, *DERIVE* will treat it like any other variable such as x or a. To verify that $\hat{e}$ is the correct number, **Author** [Alt e] and then **approX**. You should see 2.71828.

DERIVE's syntax for $\log_b x$ is `LOG(x,b)`. Thus to calculate $\log_{10} 100$, **Author** and **Simplify** `LOG(100,10)`. *DERIVE* gives the answer, 2. In *DERIVE* both `LOG(x)` and `LN(x)` refer to the *natural logarithm* or logarithm to the base e, whereas many textbooks use log to denote the logarithm to the base 10 (or *common logarithm*) and ln to denote the natural logarithm. You should also be aware that when *DERIVE* simplifies any logarithmic function, it first converts to natural logarithms, and the result may not be what you expect to see. For example, if you **Author** and **Simplify** `LOG(x,10)`, you will see $\dfrac{\text{LN}(x)}{\text{LN}(10)}$. *DERIVE* is using the formula $\log_b x = \dfrac{\log_a x}{\log_a b}$, which is true for any positive numbers, a, $b \neq 1$, and $x > 0$, with e in place of a.

Solved Problem 7.2: Logarithms, graphs, and Riemann sums

(a) Plot the graphs of $\ln x$ and $\ln(2x)$ and explain how these graphs support the logarithmic identity $\ln(2x) = \ln x + \ln 2$.

(b) Use Riemann sums to approximate $\ln 3$ with error at most 0.001. Compare your answer with *DERIVE's* approximation of $\ln 3$.

Solution to (a): **Author** and **Plot** the expressions `LNx` and `LN(2x)`. The graphs are in window 2 of Figure 7.2, and it appears that the graph of $\ln(2x)$ can be obtained by shifting the graph of $\ln x$ up by a constant. We know that $\ln(2x) = \ln 2 + \ln x$, and so this constant is in fact $\ln 2$. If we **Author** `LN2+LNx` as in expression 3 and **Plot**, we see that its graph overlays the graph of $\ln(2x)$ indicating that the two functions are identical.

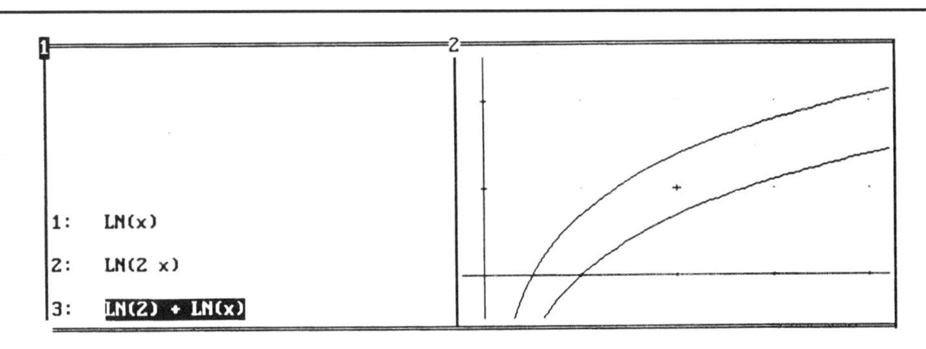

Figure 7.2: **The graphs of** $\ln x$ **and** $\ln 2x$

Solution to (b): It is a fact that $\int_1^x \frac{1}{x}\,dx = \ln x$, and some texts use this as the definition of the logarithm. Thus the value of $\ln 3$ can be approximated by approximating $\int_1^3 \frac{1}{x}\,dx$ with a right-hand sum. Furthermore, $f(x) = \frac{1}{x}$ is an increasing function, so we can use the formula

$$\text{Error } \leq |f(3) - f(1)|\Delta x$$

Thus we **Author** `F(x):=1/x`. $\Delta x = \frac{3-1}{n}$, so we next **Author** `|F(3)-F(1)|(2/n)<0.001`. When we **soLve** this inequality, we see from expression 5 of Figure 7.3 that we should choose n to be 1334. Thus, the right-hand sum we require is $\sum_{i=1}^{1334} f\left(1 + i\frac{2}{1334}\right)\frac{2}{1334}$.

1: $F(x) := \frac{1}{x}$

2: $|F(3) - F(1)| \frac{2}{n} < 0.001$

3: $n < 0$

4: $n > \frac{4000}{3}$

5: $n > 1333.33$

6: $F\left[1 + i \frac{2}{1334}\right] \frac{2}{1334}$

7: $\sum_{i=1}^{1334} F\left[1 + i \frac{2}{1334}\right] \frac{2}{1334}$

8: 1.09811

9: LN(3)

10: 1.09861

11: 1.09861 - 1.09811

12: $5.00016\ 10^{-4}$

Figure 7.3: **Approximating** $\ln 3$

To make the sum, we **Author** `F(1+i(2/1334))(2/1334)` and then use **Calculus Sum** setting Lower limit:`1` Upper limit:`1334`. When we **approX** we get the value 1.09811 in expression 8.

To get *DERIVE's* approximation for the value of $\ln 3$, we **Author** and **approX** `LN3`. In expressions 11 and 12 of Figure 7.3 we have calculated the difference between our approximation and that of *DERIVE*. We see that the actual error is 0.000500016, which is less than 0.001 as required.

Solved Problem 7.3: Approximating the number e

Plot the graph of $\left(1 + \frac{1}{n}\right)^n$ and find the smallest integer n so that $\left(1 + \frac{1}{n}\right)^n$ approximates e with error less than 0.005.

Solution We will first see *DERIVE's* approximation of e. **Author** Alt e and **approX**. The result is in expression 2 of Figure 7.4. In window 2 we have plotted the horizontal line $y = e$ along with $\left(1 + \frac{1}{n}\right)^n$. These graphs show that as n increases, the function $\left(1 + \frac{1}{n}\right)^n$ increases toward e. Thus in order to solve the problem we have to solve the equation $\left(1 + \frac{1}{n}\right)^n = e - 0.005$.

We want an estimate for n that we can use to **soLve** this equation in the approximate mode. **Author** `#1-#3-0.005` and **Plot** to get the graph in window 3 of Figure 7.4. (The **Scale** there is x:100 y:0.005.) We see that the value of n is somewhere between 200 and 300. Thus we change to the approximate mode and **soLve** expression 4 on the range Lower:200 Upper:300. From the solution in expression 5 and the requirement that n be an integer, we conclude that $n = 271$.

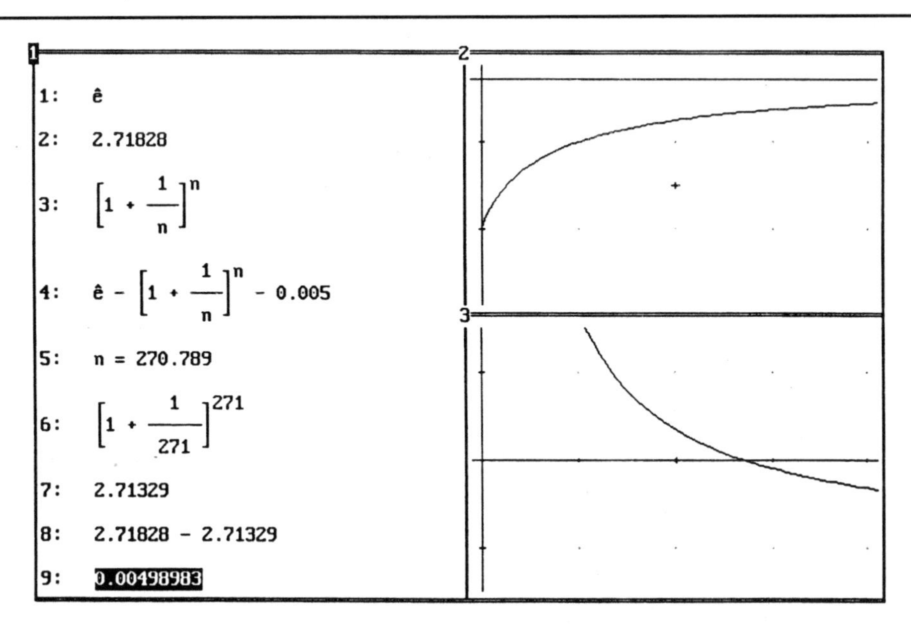

Figure 7.4: **Approximating the value of e. Scale in window 3 x:100 y:0.005**

We used **Manage Substitute** to plug $n = 271$ into expression 3 to get the approximation in expression 7 of Figure 7.4. In expressions 8 and 9 we have calculated the actual error, and it is less than 0.005 as required.

Practice Problems

1. Plot the graph of $f^{-1}(x)$ if $f(x) = x^3 + x$.

2. Use *DERIVE* to calculate $\log_6 8$. Answer: 1.16055

3. Solve $\ln(e^x + 2) = 7$. Answer: $\ln(e^7 - 2)$

4. Use *DERIVE* to calculate $\frac{d}{dx} e^x \ln(1 + e^x)$. Answer: $\dfrac{e^x(e^x(\ln(e^x+1)+1)+\ln(e^x+1))}{e^x+1}$

5. Use *DERIVE* to calculate $\int e^x \cos x \, dx$. Answer: $e^x \left(\dfrac{\cos x + \sin x}{2}\right)$

Laboratory Exercise 7.1

Inverses of Functions

Name ______________________________ Due Date ____________

For each of the following functions

(a) Plot the graph of f and determine if it has an inverse. (Your determination should include both graphical evidence and an analytical justification.)

(b) If the function fails the horizontal line test, include the graph of a horizontal line that shows this.

(c) If the function has an inverse, plot its graph.

(d) If the function has an inverse, find $f^{-1}(2)$.

1. $f(x) = x^3 + 3x - 5$

2. $f(x) = \dfrac{x^2+1}{x^4+1}$

3. $f(x) = x^3 + \sin x$

4. $f(x) = \sqrt{x^4 + x + 1}$

Laboratory Exercise 7.2

Restricting the Domain

Name ______________________________ Due Date ____________

If we restrict the domain of a function, we can make it pass the horizontal line test for invertibility. For example, $f(x) = x^2$ is not invertible, but $f(x) = x^2$ restricted to $[0, \infty)$ is invertible, and $f^{-1}(x) = \sqrt{x}$.

1. $g(x) = x^2$ restricted to $(-\infty, 0]$ is invertible. What is $g^{-1}(x)$?

2. Show that $h(x) = x^3 + x^2 + 1$ is not invertible.

3. Find the largest interval of the form (a, ∞) on which $h(x) = x^3 + x^2 + 1$ is invertible. Explain how you got your answer.

4. Plot the graph of the inverse you found in Part 3.

5. Find two other intervals on which $h(x)$ is invertible and plot the graphs of the inverses.

Laboratory Exercise 7.3

Seeing Log Identities Graphically

Name ______________________________ **Due Date** ____________

1. Plot the graphs of $\ln x$ and $\ln(3x)$ and explain how these graphs support the identity $\ln(3x) = \ln 3 + \ln x$.

2. Plot the graphs of $\ln x$ and $\ln(x^2)$ and explain how these graphs support the identity $\ln(x^2) = 2 \ln x$.

3. Plot the graphs of $\ln x$ and $\ln\left(\frac{x}{2}\right)$ and explain how these support the identity $\ln\left(\frac{x}{2}\right) = \ln x - \ln 2$.

4. Plot $y = x^2$ and $y = e^{2 \ln x}$. What do you observe? What does it tell you about the two functions?

Laboratory Exercise 7.4

Comparing Logarithms to Different Bases

Name __________________________________ **Due Date** ____________

1. On the same screen, plot the graphs of $\ln_2 x$, $\ln_3 x$, $\ln_4 x$, and $\ln_5 x$.

2. Explain how the size of the positive base affects the logarithm. Prove any assertions you make.

3. On the same screen, plot the graphs of $\ln_2 x$ and $\ln_{\frac{1}{2}} x$.

4. Explain the relationship between $\ln_b x$ and $\ln_{\frac{1}{b}} x$ for positive b. Prove any assertions you make.

Laboratory Exercise 7.5

Growth Rates of Functions

Name ______________________________ **Due Date** ____________

1. On the same screen, plot the graphs of $\ln x$, $x^{\frac{1}{2}}$, x, x^2, and e^x.

2. Explain the relationships among these five functions for large values of x.

3. Based on your comparison, what do you think is the value of $\lim_{x \to \infty} \frac{x^{100}}{e^x}$? Provide graphical evidence for your assertion and verify it by asking *DERIVE* to calculate the limit.

4. Based on your comparison from Part 2, what do you think is the value of $\lim_{x \to \infty} \frac{\ln x}{\sqrt{x}}$? Provide graphical evidence for your assertion and verify it by asking *DERIVE* to calculate the limit.

Chapter 8

Hyperbolic and Inverse Trigonometric Functions

New *DERIVE* topics •Syntax for hyperbolic and inverse trig functions
Calculus concepts •Hyperbolic functions •Inverse trigonometric functions •Inverse hyperbolic functions

To indicate the hyperbolic functions *DERIVE* puts an `H` at the end of the standard syntax for the six trigonometric functions. Thus the syntax for the hyperbolic sine is `SINH`, and the syntax for the hyperbolic cosine is `COSH`. To indicate the inverse of both trigonometric and hyperbolic functions, *DERIVE* puts an `A` at the front. Thus the syntax for the inverse tangent function is `ATAN`, and the syntax for the inverse hyperbolic sine is `ASINH`.

Solved Problem 8.1: Arc length and area with inverse trig functions

(a) Find the arc length of $\arctan x$ from $x = 0$ to $x = 1$.

(b) Find the area of the "triangular region" enclosed by the graphs of $\arctan x$, $\arccos x$, and the y-axis.

Solution to (a): First **Author** `ATANx` and then `SQRT(1+DIF(#1,x)^2)`. The simplified form of this is in expression 3 of Figure 8.1. We use **Calculus Integrate** setting Lower limit:`0` Upper limit:`1`, and then **approX** to get the result in expression 5.

In window 3 we have plotted the graphs to get a look at the region. We see that in the region of interest, the graph of $\arccos x$ is on top, and we need to find the x-coordinate of the point where they cross. To do this we **Author** `#6-#1`, change to the approximate mode using **Options Precision Approximate**, and **soLve** expression 7 on the range Lower:`0` Upper:`1`. The result is in expression 8. To get the area we highlight expression 7 and use **Calculus Integrate** with Lower limit:`0` Upper limit:`0.786151`. We **approX** to get the area in expression 10.

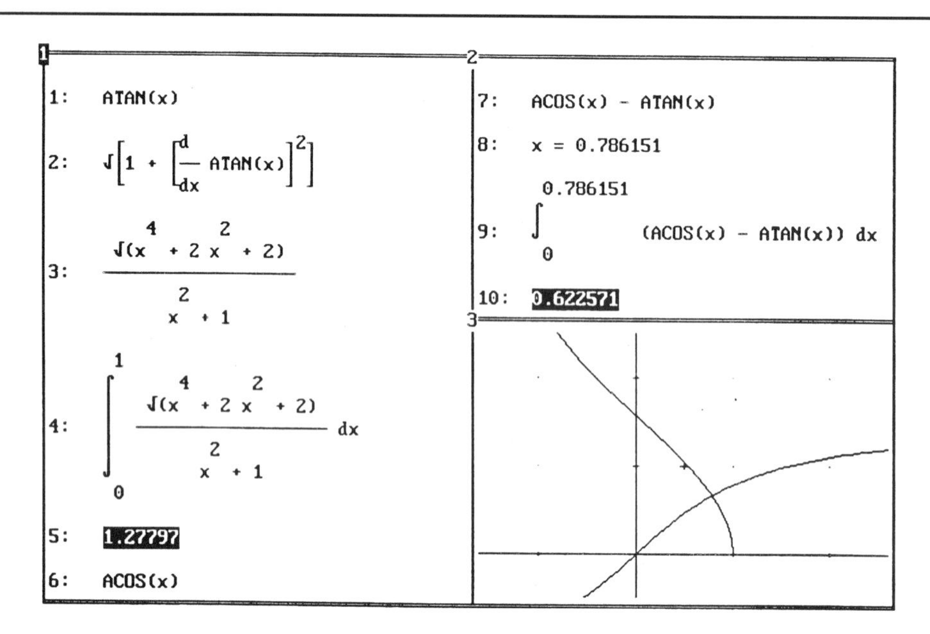

Figure 8.1: **Arc length and area with** $\arctan x$ **and** $\arccos x$

Practice Problems

1. Use *DERIVE* to calculate $\dfrac{d}{dx}(\arctan x \sinh x)$. Answer: $\dfrac{e^x + e^{-x}}{2} \arctan x + \dfrac{e^{-x}(e^{2x} - 1)}{2(x^2 + 1)}$

2. Use *DERIVE* to calculate the antiderivative of the inverse hyperbolic sine function. Answer: $x \ln(\sqrt{x^2 + 1} + x) - \sqrt{x^2 + 1}$

Laboratory Exercise 8.1

Lengths, Areas, and Volumes with Hyperbolic Functions

Name ______________________________ **Due Date** ____________

Let R denote the region enclosed by $f(x) = 2 - \cosh x$ and the x-axis.

1. Find the area of R.

2. Find the "perimeter" of R.

3. Find the volume of the solid of revolution obtained by rotating R around the x-axis.

4. Find the surface area of the solid in Part 3.

Laboratory Exercise 8.2

A Hanging Chain

Name ______________________________ **Due Date** __________

If the ends of a chain are attached to the points $(-1, 0)$, and $(1, 0)$, it will take the shape of

$$\frac{\cosh(ax) - \cosh a}{a}$$

on the interval $[-1, 1]$, where a depends on the length of the chain.

1. Find the length of the chain as a function of a.

2. Find the value of a for which the chain has length 3. Plot the graph.

3. For what value of a does the chain hang 1 unit below the x-axis at its lowest point? Explain how you got your answer.

4. Find the length of the chain using the value of a you found in Part 3.

Chapter 9

Numerical Integration

New *DERIVE* topics
Calculus concepts •The trapezoidal rule •The midpoint rule •Simpson's rule •Error in Simpson's rule and the trapezoidal rule

In Chapter 5 we used left-hand and right-hand Riemann sums to approximate integrals; however, for many integrals that cannot be evaluated directly using the Fundamental Theorem of Calculus these methods are not fast or accurate enough for practical application. The approximation techniques presented here are more often used, and you may be interested to know that *DERIVE* uses a form of one of them, Simpson's rule, to make its approximations.

Solved Problem 9.1: Implementing the left-hand rule and right-hand rule

Create a file that automates the left-hand rule and right-hand rule. Use it to estimate $\int_1^4 \sin x \, dx$.

Solution: We will make a function RIGHT(a,b,n) (See Figure 9.1) which calculates the right-hand Riemann sum for f on the interval $[a, b]$ using n subintervals. It is tested in Figure 9.2. The sum we want is $\sum_{i=1}^{n} f(a+i\Delta x)\Delta x$ where $\Delta x = \frac{b-a}{n}$. To do this, begin with a clear screen and **Author** the following as expressions 1, 2, and 3.

```
F(x):=

(b-a)/n

F(a+i #2) #2
```

Now, with the third expression highlighted, use **Calculus Sum** setting Calculus Sum variable: `i` and Lower limit:`1` Upper limit:`n`. The correct sum is in expression 4 of Figure 9.1. To finish **Author** `RIGHT(a,b,n):=#4`.

To make the LEFT function, highlight expression 3 once more and use **Calculus Sum**, but this time set the limits to Lower limit:`0` Upper limit:`n-1`. The sum is in expression 6 of Figure 9.1. To complete the LEFT function, **Author** `LEFT(a,b,n):=#6`. You can save this file using **Transfer Save Derive** SUMS and call it back when you want using **Transfer Merge** SUMS.

1: $F(x) :=$

2: $\dfrac{b - a}{n}$

3: $F\left[a + i\,\dfrac{b - a}{n}\right] \dfrac{b - a}{n}$

4: $\displaystyle\sum_{i=1}^{n} F\left[a + i\,\dfrac{b - a}{n}\right] \dfrac{b - a}{n}$

5: $\displaystyle \text{RIGHT}(a, b, n) := \sum_{i=1}^{n} F\left[a + i\,\dfrac{b - a}{n}\right] \dfrac{b - a}{n}$

6: $\displaystyle\sum_{i=0}^{n-1} F\left[a + i\,\dfrac{b - a}{n}\right] \dfrac{b - a}{n}$

7: $\displaystyle \text{LEFT}(a, b, n) := \sum_{i=0}^{n-1} F\left[a + i\,\dfrac{b - a}{n}\right] \dfrac{b - a}{n}$

Figure 9.1: **The left-hand rule and the right-hand rule**

```
8:  F(x) := SIN(x)

9:  RIGHT(1, 4, 10)

10: 0.945237

11: LEFT(1, 4, 10)

12: 1.42471
```

Figure 9.2: **Testing the RIGHT and LEFT functions**

To use the RIGHT function to approximate $\int_1^4 \sin x \, dx$ with 10 subintervals, we first **Author** `F(x):=SINx` as seen in expression 9 of Figure 9.2. Now **Author** `RIGHT(1,4,10)` and **approX**. To use the LEFT function for the same integral, **Author** `LEFT(1,4,10)` and **approX**. The result is in expression 12.

Solved Problem 9.2: Making pictures of approximation techniques

Use the RIEMANN.MTH and the SIMPSON.MTH file from Appendix II to make pictures of the left-hand sum, the right-hand sum, the midpoint rule, the trapezoidal rule, and Simpson's rule for $\int_0^2 \sin x \, dx$ using four subintervals.

Solution: We have already used the RIEMANN.MTH file to make pictures of left-hand sums and right-hand sums, but as we see in Figure 9.3, the file can also make pictures of the midpoint rule and of the trapezoidal rule. The SIMPSON.MTH file from Appendix II is required to make the picture of Simpson's rule in Figure 9.4.

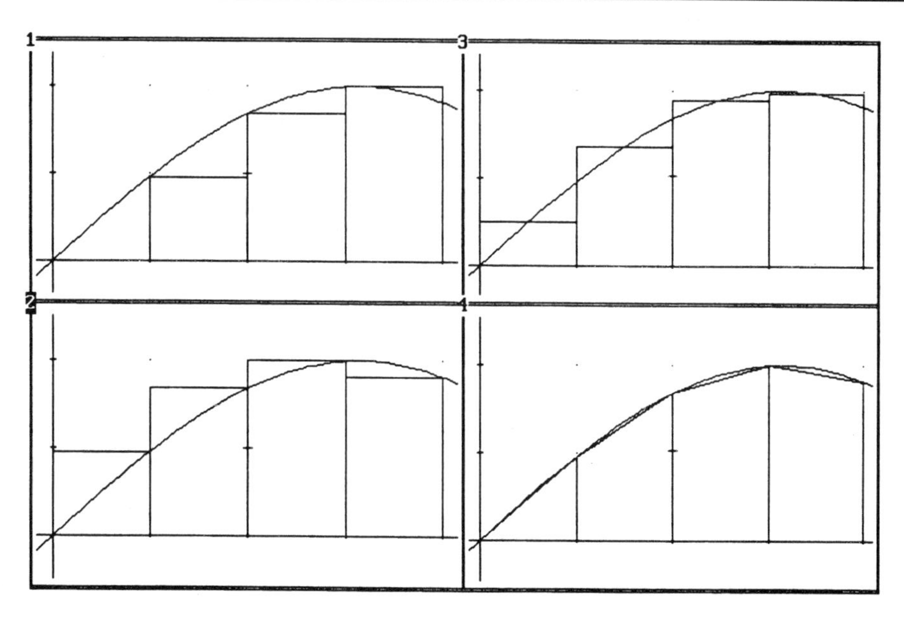

Figure 9.3: **The left-hand sum, right-hand sum, midpoint rule, and trapezoidal rule. Scale x:0.5 y:0.5**

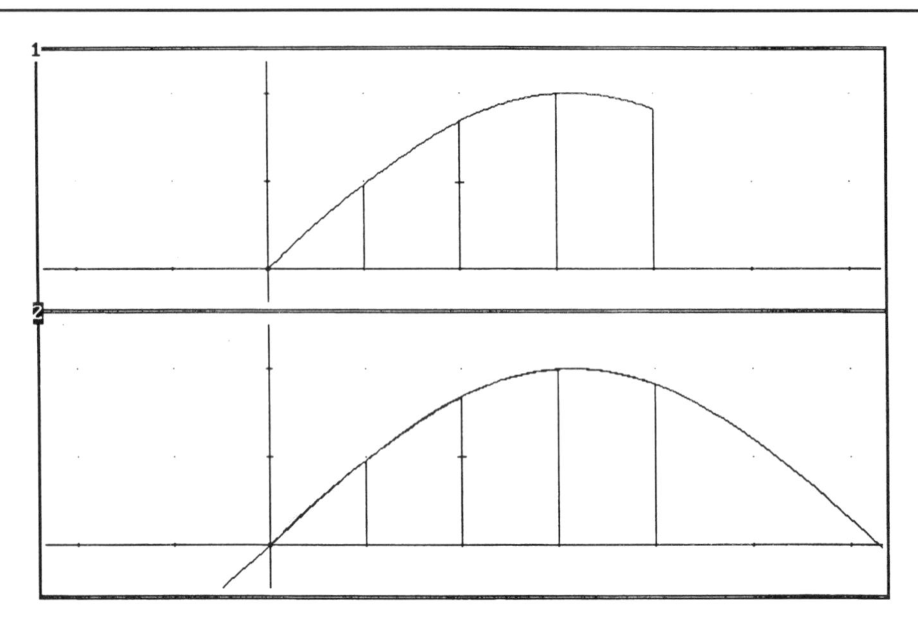

Figure 9.4: **A picture of Simpson's rule. Scale x:0.5 y:0.5**

Load the file using **Transfer Merge** RIEMANN. Now **Author** `F(x):=SINx` to tell *DERIVE* which function we want. To make the left-hand sum **Author** and **approX** `LPIC(0, 2, 4)`. When you **Plot**, don't forget to change the graphics settings using **Options State Rectangular Connected Small**. **Plot** again, and you will see the picture in window 1 of Figure 9.3. In window 2 we have used `RPIC(0, 2, 4)` to make a picture of the right-hand sum. In window 3 we used `MPIC(0, 2, 4)` to make a picture of the midpoint rule, and in window 4, we used `TPIC(0, 2, 4)` to make a picture of the trapezoidal rule. In each case we used F9 to zoom in once. If you look carefully at each of the pictures, you should be able to explain where each of these four approximation methods got their names.

Load the SIMPSON.MTH file using **Transfer Merge** SIMPSON. As with the earlier pictures, we **Author** `F(x):=SINx` to tell *DERIVE* which function we want. Now **Author** and **approX** `SPIC(0,2,4)`. When you ask *DERIVE* to **Plot**, you will be prompted with PLOT: Min:-3.1416 Max:3.1415. (The plot is parametric.) Change this to Min:0 Max:1. (Use the Tab key to set the Max: field.) The picture appears in window 1 of Figure 9.4.

Since the graph of the sine function is not automatically added by this file, we have included it in window 2. Simpson's rule is approximating the function value over each interval with a quadratic. It is difficult to distinguish the approximations from the sine function in window 2, but you can when you see it in color.

Solved Problem 9.3: The trapezoidal rule with error control

If the trapezoidal rule is used to approximate $\int_a^b f(x)\,dx$ using n subintervals, then the error in the approximation is no more than

$$\text{Trapezoidal error} \le \frac{K(b-a)^3}{12n^2}$$

where K is the maximum of $|f''(x)|$ on the interval $[a, b]$.

Find a value of n that assures that the error in using the trapezoidal rule to approximate $\int_0^2 \sqrt{1+x^4}\,dx$ is no more than 0.001. Calculate the approximation using the value of n that you found.

Solution: Before working through this Solved Problem, you will find it helpful to complete Laboratory Exercise 9.1. Load the SUMS.MTH file you make there using **Transfer Load Derive** SUMS. Now **Author** `F(x):=SQRT(1+x^4)` to tell *DERIVE* which function we want. The first step is to find the value of K. To do that we look at the second derivative. With expression 14 of Figure 9.5 highlighted, use **Calculus Differentiate** setting Order:2. **Simplify** to get the second derivative which we have plotted in window 3. From the picture, we see that the absolute value of the second derivative is never larger than 3 on the interval $[0, 2]$, and so we take $K = 3$. You should be able to use your knowledge of calculus to find the exact value of the maximum here, and there are some advantages to doing so as you will discover in Laboratory Exercise 9.3. But all we really need is a number that we are sure is larger than $|f''(x)|$ over the interval $[0, 2]$, so we just estimate it from the graph. Thus, the inequality we need to solve is

$$\frac{3(2-0)^3}{12n^2} \le 0.001$$

Author `(3 2^3 )/(12n^2)<0.001` and **soLve**. From expression 20 of Figure 9.5 we see that $n = 45$ will work. In expressions 21 and 22 we have calculated the trapezoidal approximation using this value of n. *DERIVE's* approximation of this integral appears in expression 25, and you can verify that it indeed differs from our approximation by less than 0.001.

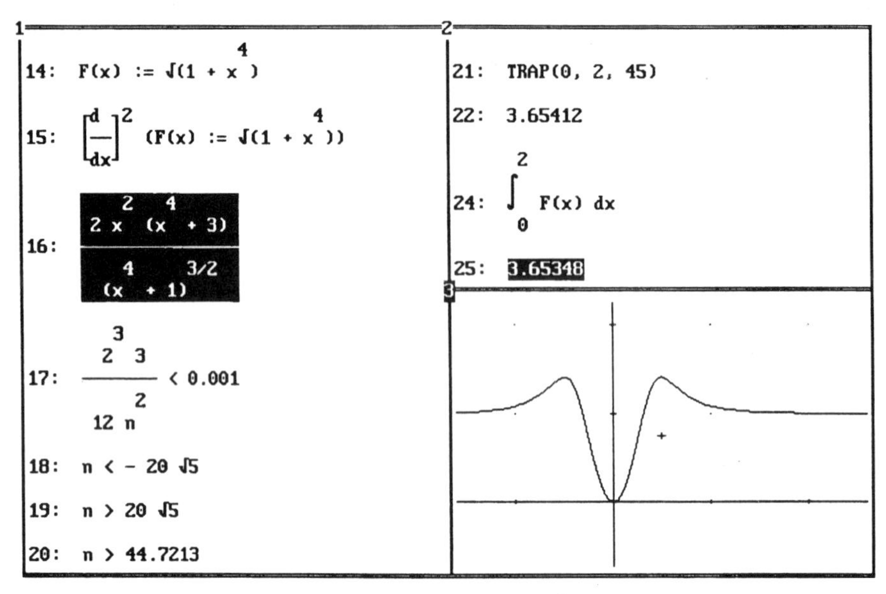

Figure 9.5: **Trapezoidal approximation for** $\int_0^1 (1+x^4)^{\frac{1}{2}}\,dx$**, Scale x:2 y:2**

Practice Problems

1. Use the LEFT and RIGHT functions to approximate $\int_0^2 \cos x\,dx$ using $n = 10$. Compare these with the exact value. <u>Answer</u>: LEFT$(0,2,10) = 1.04787$ RIGHT$(0,2,10) = 0.764649$

Laboratory Exercise 9.1

Completing the SUMS.MTH File

Name ______________________________ **Due Date** __________

Begin this exercise by using **Transfer Load Derive** SUMS.MTH to recall the file you constructed in Solved Problem 9.1. (Respond with `Y` when prompted with Abandon expressions (Y/N)?.)

1. Create a function TRAP(a, b, n) that implements the trapezoidal rule. (The *trapezoidal rule* is the average of the left-hand rule and the right-hand rule.)

2. Create a function MID(a, b, n) that implements the midpoint rule. The midpoint rule is a Riemann sum with equally spaced subintervals that uses the midpoint of each subinterval. (Suggestion: You may edit expression 3 of Figure 9.1 by using **Manage Substitute** to replace i by `i-1/2`. Then use **Calculus Sum**.)

3. Create a function SIMP(a, b, n) that implements Simpson's rule. (Simpson's rule is a weighted average of the midpoint rule and the trapezoidal rule.) We remark that this is what many books would call Simpson's rule for $2n$ subintervals.

$$\text{SIMP}(a, b, n) = \frac{2\text{MID}(a, b, n) + \text{TRAP}(a, b, n)}{3}$$

4. Create a function ALL(a, b, n) that calculates all five sums at once as follows:

```
ALL(a,b,n):=[LEFT(a,b,n), RIGHT(a,b,n), MID(a,b,n), TRAP(a,b,n), SIMP(a,b,n)]
```

5. Save your completed work using **Transfer Save Derive** SUMS.

Laboratory Exercise 9.2

Comparing Approximation Techniques

Name ______________________________ Due Date ____________

Consider $\int_0^1 \frac{1}{1+x^2}\,dx$.

1. Use the Fundamental Theorem of Calculus to show that the value of the integral is $\frac{\pi}{4}$.

2. Use the ALL function that you made in Laboratory Exercise 9.1 to obtain approximations of the integral using a left-hand sum, a right-hand sum, the midpoint rule, the trapezoidal rule, and Simpson's rule with $n = 4$. Compare the results with the exact value of the integral.

3. Use the RIEMANN.MTH and SIMPSON.MTH files from Appendix II to make pictures of each of the five approximations you made in Part 1.

Laboratory Exercise 9.3

The Trapezoidal Rule and Simpson's Rule with Error Control

Name ______________________________ Due Date ____________

This laboratory is a continuation of Solved Problem 9.3.

1. Find the maximum of $|f''(x)|$ on $[0, 2]$ and explain how you got your answer.

2. In the solution of Solved Problem 9.3 we found a number K that was larger than $|f''(x)|$ on the interval $[0, 2]$. Explain why your answer in Part 1 may also be used as a value of K.

3. What value for n do you get in the error formula for the trapezoidal rule if you use the value of K you found in Part 1?

4. Find the trapezoidal approximation using the value of n you got in Part 3.

5. If Simpson's rule is used to approximate $\int_0^1 f(x)\, dx$ using n subintervals, then the error is no more than

$$\frac{M(b-a)^5}{180n^4}$$

where M is the maximum value of $|f^{(4)}(x)|$ on the interval $[a, b]$. Find a value of n so that Simpson's rule approximates $\int_0^2 \sqrt{1+x^4}\, dx$ with error no more than 0.001. Explain how you got the value of M. Calculate Simpson's rule using this value of n and compare your answer with *DERIVE's* approximation of the integral.

Chapter 10

Improper Integrals

New *DERIVE* topics •Evaluating improper integrals •Approximating improper integrals Calculus concepts •Improper integrals •The comparison theorem

Solved Problem 10.1: Calculating improper integrals

New *DERIVE* Lessons •Evaluating improper integrals

Calculate each of the following integrals by using *DERIVE* to get the answer directly and by expressing it as a limit of a Riemann integral.

(a) $\int_0^\infty e^{-\sqrt{x}}\,dx$ (b) $\int_0^3 \frac{1}{(x-2)^2}\,dx$ (c) $\int_0^\infty \cos^3 x\,dx$

Solution to (a): First **Author** Alt e `^-SQRT(x)`. Now use **Calculus Integrate** setting Lower limit:0 Upper limit:`inf`. **Simplify** to get the answer, 2, in expression 3 of Figure 10.1. The definition of the improper integral is $\int_0^\infty e^{-\sqrt{x}}\,dx = \lim_{k\to\infty}\int_0^k e^{-\sqrt{x}}\,dx$. Highlight expression 1 and use **Calculus Integrate** once more. This time set Lower limit:0 Upper limit:`k`. **Simplify** to get the answer in expression 5 of Figure 10.1. We want the limit of this as k goes to infinity. Thus we use **Calculus Limit** setting Point:`inf`. The limit appears in expression 6, and when we **Simplify**, we see from expression 7 that the answer agrees with *DERIVE's* earlier calculation.

Solution to (b): **Author** `1/(x-2)^2` and use **Calculus Integrate** with Lower limit:0 Upper limit:3. When we **Simplify**, *DERIVE* returns the answer $-\frac{3}{2}$ in expression 10 of Figure 10.1. But this can't be right; the integrand is never negative, and so the integral cannot be negative.

When we asked *DERIVE* to **Simplify**, it found the antiderivative of $\frac{1}{(x-2)^2}$ and applied the Fundamental Theorem of Calculus to get $\left.\frac{1}{2-x}\right|_0^3 = -\frac{1}{2} - 1 = -\frac{3}{2}$. Because the function f is not continuous at $x = 2$, this is not a valid application of the Fundamental Theorem of Calculus. The *DERIVE* User Manual warns that this is the method that is used, and you must take care. The moral is clear: *DERIVE* is a highly efficient calculating tool, but it cannot replace your ability to think.

This integral exists provided *both* $\int_0^2 \frac{1}{(x-2)^2}\,dx$ and $\int_2^3 \frac{1}{(x-2)^2}\,dx$ converge. In expressions 11 and 12 of Figure 10.1 we have calculated the first of these integrals, and *DERIVE*

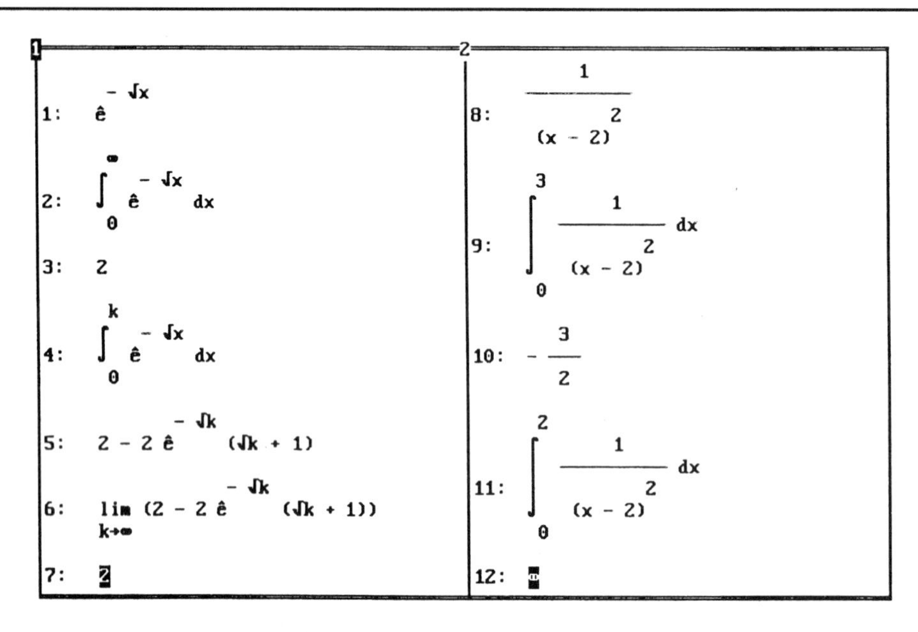

Figure 10.1: **Evaluating improper integrals**

correctly reports that it diverges. It is not necessary to check the second integral; we conclude that $\int_0^3 \frac{1}{(x-2)^2}\,dx$ does not exist.

Solution to (c): **Author** `COS^3x` (not shown in Figure 10.2) and ask *DERIVE* to calculate the integral as in Part (a). We see in expression 3 that *DERIVE* returns the meaningless expression $\sin(\infty)$. In expressions 4 through 7 we calculated the integral as a limit of a Riemann integral and got the same answer. We need more information to determine the value of the integral. In window 2 of Figure 10.2 we have plotted $\int_0^k \cos^3 x\,dx$ as a function of k (expression 5). We see from the graph that as k increases, the antiderivative oscillates and does not appear to approach a single value. You should be able to verify this analytically, and we conclude that the integral does not exist.

Solved Problem 10.2: Approximating improper integrals

New *DERIVE* Lessons	•Approximating improper integrals

Determine if the following integrals exist. If they do, approximate their values.

(a) $\int_2^{\infty} \frac{1}{\ln x}\,dx$ (b) $\int_2^{\infty} \frac{1}{x^2+\ln x}\,dx$

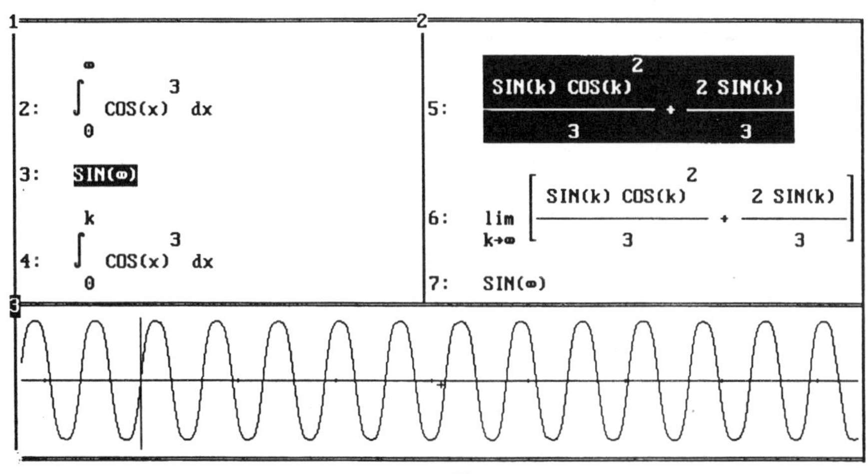

Figure 10.2: **Divergence of** $\int_0^\infty \cos^3 x\, dx$, **Scale x:10 y:1**

Solution to (a): **Author** `1/LNx` and use **Calculus Integrate** setting Lower limit:2 Upper limit:`inf`. If we ask *DERIVE* to **Simplify**, it returns the integral unevaluated in expression 3 of Figure 10.3. *DERIVE* cannot evaluate the integral exactly, but it can approximate improper integrals just as it does Riemann integrals. When we **approX** expression 3, *DERIVE* returns a number in expression 4, but not before beeping and displaying the message "Dubious accuracy" in the lower left corner of the screen. *DERIVE* is telling us that it is uncertain about the accuracy of the result. This should cause us to wonder if the integral converges at all. How can we tell? We will make use of the following theorem which may be in your text.

Comparison theorem for improper integrals: Let f and g be continuous on the open interval (a, b), and suppose $0 < f(x) < g(x)$. If $\int_a^b g(x)\, dx$ converges, then so does $\int_a^b f(x)\, dx$. If $\int_a^b f(x)\, dx$ diverges, then so does $\int_a^b g(x)\, dx$.

Intuitively, the comparison theorem says that integrals that are smaller than convergent integrals converge, and integrals that are larger than divergent integrals diverge. In the case we are dealing with, we note that since $0 < \ln x < x$, we have $\dfrac{1}{\ln x} > \dfrac{1}{x}$ on the interval $(2, \infty)$. We can determine by direct calculation that $\int_2^\infty \frac{1}{x}\, dx$ diverges, so we conclude that our integral diverges also. What of *DERIVE's* answer in expression 4? It's wrong, and *DERIVE* was smart enough to warn us before giving the answer. In general, it is good practice to establish

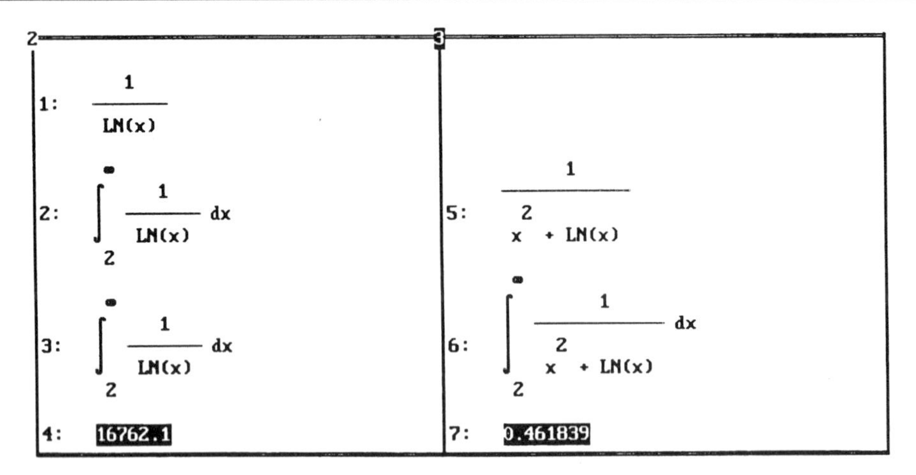

Figure 10.3: **Approximating improper integrals**

convergence before asking *DERIVE*, or any other computer program, for an approximate answer.

Solution to (b): Profiting from the lesson we learned in Part (a), we note that $\frac{1}{x^2 + \ln x} < \frac{1}{x^2}$ on the interval $(2, \infty)$ and that $\int_2^\infty \frac{1}{x^2}\, dx$ converges. From the comparison theorem, we conclude that our integral also converges. When we **approX** this integral, it returns the answer in expression 7 of Figure 10.3, and this time we do not see the "Dubious accuracy" warning.

Practice Problems

1. Use *DERIVE* to calculate $\int_0^\infty x e^{-\sqrt{x}}\, dx$. Answer: 12

2. Use *DERIVE* to calculate $\int_0^\infty \frac{1}{1+\sqrt{x}}\, dx$. Answer: ∞

3. Approximate the value of $\int_0^\infty \frac{1}{x^2 + \sin x + 2}\, dx$. Answer: 0.995713

Laboratory Exercise 10.1

Evaluating Improper Integrals

Name ______________________________ **Due Date** ____________

Calculate the following integrals. Verify your answers by calculating the appropriate limit. Include a graph of each integrand and give a geometric interpretation of the integral.

1. $\displaystyle\int_0^{\infty} e^{-2x} \cos x \, dx$

2. $\displaystyle\int_0^{1} \ln x \, dx$

3. $\displaystyle\int_0^{3} \frac{1}{x^2 + x - 1} \, dx$

4. $\displaystyle\int_0^3 \frac{1}{\sqrt{|x^2+x-1|}}\,dx$ (Hint: Decide where x^2+x-1 is positive and where it is negative.)

5. $\displaystyle\int_{-\infty}^{\infty} x \sin x \, dx$

Laboratory Exercise 10.2

Applications of Improper Integrals

Name ______________________________ Due Date __________

Solve the following problems and explain how you use improper integrals.

1. **A solid of revolution**: Let R be the solid of revolution determined by rotating the graph of $\frac{1}{x}$ from $x = 1$ to infinity around the x-axis. (a) Find the volume of R. (b) Find the surface area of R.

2. **A rocket ship**: This is a continuation of Part 3 of Laboratory Exercise 2.3. Find the work required to move the rocket ship from the surface of the Earth to a point infinitely far away. (Hint: The gravitational force acting on the rocket when it is a distance d from the center of the Earth is $G\frac{M_e M_d}{d^2}$ where $M_e = 5.98 \times 10^{24}$ kilograms is the mass of the Earth, M_d is the mass of the rocket given in Laboratory Exercise 2.3, and $G = 6.67 \times 10^{-11}$ is the gravitational constant.)

3. **The Laplace transform**: For a function $f(x)$, the *Laplace transform* of f is defined to be $L(s) = \int_0^\infty e^{-st} f(t)\, dt$ if the integral converges. The Laplace transform is important for the solution of certain types of differential equations. Find the Laplace transform of $\sin x$. (Note: *DERIVE* cannot calculate the integral unless you first **Declare Variable s Positive**. Your final answer should be a function of s.)

4. The Gamma function, which we introduced in Laboratory Exercise 3.5, is defined by

$$\Gamma(x) = \int_0^\infty t^{x-1} e^{-t}\, dx$$

(a) Calculate $\Gamma\left(\frac{3}{2}\right)$ using the above definition.

(b) Check your answer in Part (a) by asking *DERIVE* to calculate $\Gamma\left(\frac{3}{2}\right)$ directly. (Note: *DERIVE's* syntax for the Gamma function is [Alt G](x).)

Laboratory Exercise 10.3

Approximating Improper Integrals

Name ______________________ **Due Date** __________

Use the comparison theorem to determine if the following integrals converge or diverge. For those that converge, use *DERIVE* to approximate them.

1. $\int_1^\infty \frac{1}{x+e^x}\,dx$ (Suggested comparison function: e^{-x})

2. $\int_2^\infty \frac{1}{x+\ln x}\,dx$. (Suggested comparison function: $\frac{1}{2x}$)

3. $\int_0^1 \frac{-\ln x}{1+x}\, dx$ (Suggested comparison function: $-\ln x$)

4. $\int_2^\infty \frac{1}{\ln^2 x}\, dx$ (Suggested comparison function: $\frac{1}{x \ln x}$)

Chapter 11

Infinite Series

New *DERIVE* topics •Summing infinite series •Approximating infinite series •Plotting partial sums •Calculating Taylor polynomials
Calculus concepts •Infinite series •Convergence tests •Radius of convergence •Taylor and Maclaurin series •Remainder term

Solved Problem 11.1: Summing convergent series

New *DERIVE* Lessons •Summing infinite series •Approximating infinite series

Show that the following series converge and find or approximate their sums.

(a) $\sum_{n=1}^{\infty} \frac{1}{n^2}$ (b) $\sum_{n=1}^{\infty} \frac{1}{n^3}$ (c) $\sum_{n=1}^{\infty} \frac{1}{n^2+1}$

Solution to (a): Your text may discuss a theorem that says $\sum_{n=1}^{\infty} \frac{1}{n^p}$ converges when $p > 1$. (It follows from the integral test.) To find the sum **Author** `1/n^2` and use **Calculus Sum**, setting `Lower limit:1 Upper limit:inf`. (Remember to use the Tab key to get to the Upper limit: field.) **Simplify** to get the answer, $\frac{\pi^2}{6}$, in expression 3 of Figure 11.1.

Solution to (b): We can establish convergence using the same method we used in Part (a). This time, when we make the sum and **Simplify**, *DERIVE* returns ZETA(3). The ZETA function is important in number theory and is defined as $\zeta(p) = \sum_{n=1}^{\infty} \frac{1}{n^p}$. It is the same as the p-series, and the sum we calculated in Part (a) is in fact $\zeta(2)$. The exact value of $\zeta(3)$ is not known, but we can **approX** to get the approximation in expression 7 of Figure 11.1. It is a curious fact that the exact value of $\zeta(n)$ is known for all even integers n, but not for any odd integer greater than 1. (What is the value of $\zeta(1)$?)

Solution to (c): Since $\frac{1}{n^2+1} < \frac{1}{n^2}$, the comparison test assures us that $\sum_{1}^{\infty} \frac{1}{n^2+1}$ converges. (Alternatively, we may use the integral test.) The sum appears in expression 9 of Figure 11.1. When we **Simplify**, *DERIVE* returns the series in expression 10 unevaluated, indicating that it does not know the answer. The same thing happens if we **approX**. (Try it.)

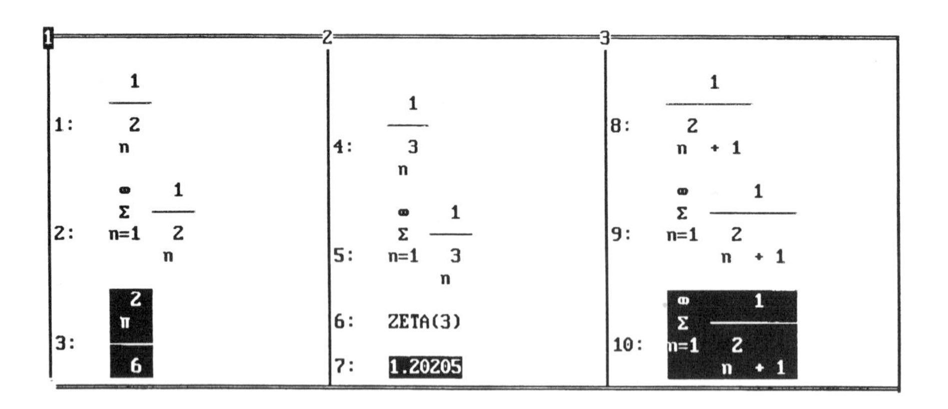

Figure 11.1: **Summing infinite series**

One way to approximate the value of the series is to calculate partial sums $\sum_{n=1}^{k} \dfrac{1}{n^2+1}$ for large values of k. Highlight expression 8 of Figure 11.1 and use **Calculus Sum** once more setting Lower limit:1 Upper limit:k. Now use **Manage Substitute** to replace k by 100 as we have done in expression 12 of Figure 11.2. When we **approX**, *DERIVE* gives the answer in expression 13.

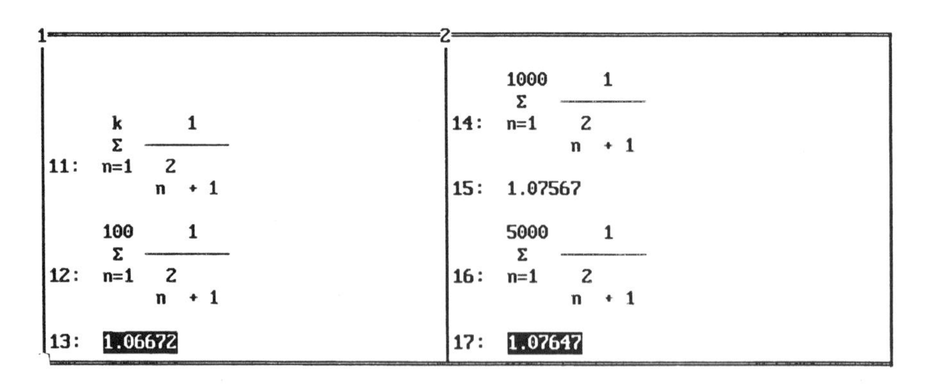

Figure 11.2: **Approximating** $\sum_{n=1}^{\infty} \dfrac{1}{n^2+1}$

In expressions 14 through 17 we have calculated the partial sums for $k = 1000$ and $k = 5000$. We don't see much change from expression 15 to expression 17, so it seems reasonable to give 1.07647 as an estimate of the sum of the series. It is important to note, however, that even though this answer has intuitive appeal, we have not established how accurate it is.

Solved Problem 11.2: The harmonic series and the logarithm

New *DERIVE* Lessons •Plotting partial sums

Plot the graph of the partial sums of the harmonic series $\sum_{n=1}^{\infty} \frac{1}{n}$ along with the graph of $\ln x$. Explain how these graphs give evidence for the divergence of the harmonic series.

Solution: **Author** `1/n` and use **Calculus Sum** setting Lower limit:`1` Upper limit:`k`. Now **Plot**, and *DERIVE* will make the graph treating k as the variable. In window 2 of Figure 11.3 we have **Center**ed in the first quadrant.

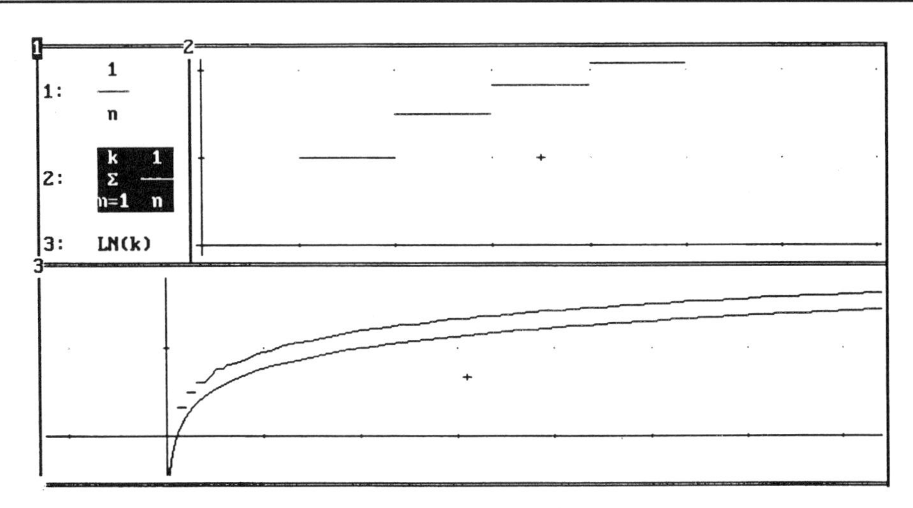

Figure 11.3: **Divergence of the harmonic series. Scale in window 3 x:10 y:3**

We are interested in what happens to the graph for large values of k, so in window 3 we have set the **Scale** to x:10 y:3 and **Center**ed once more in the first quadrant. When we do this, the graph appears to smooth out. Next we plot $\ln k$. It appears that the graph of the partial sums stays above the graph of the logarithm, and since we know that $\lim_{k\to\infty} \ln k = \infty$, the partial sums of the harmonic series must also go to infinity. You can verify this using the integral test.

You may find it interesting also to plot the graph of $1 + \ln x$. The resulting picture suggests a close connection between $\ln x$ and the partial sums of the harmonic series. This is explored further in Laboratory Exercise 11.3.

Solved Problem 11.3: Radius of convergence

It can be shown that $\dfrac{10}{x^2+9} = \sum_{n=0}^{\infty} \dfrac{10(-1)^n}{9^{n+1}} x^{2n}$ when the series converges.

(a) Plot the graphs of $\dfrac{10}{x^2+9}$ and the partial sums $\sum_{n=1}^{k} \dfrac{10(-1)^n}{9^{n+1}} x^{2n}$ for $k = 1$ through 5, and use the picture to estimate the radius of convergence.

(b) Use the ratio test to calculate the radius of convergence exactly.

Solution to (a): First **Author** `10/(x^2+9)` (not shown in Figure 11.4) and **Plot** to get the graph in window 2.

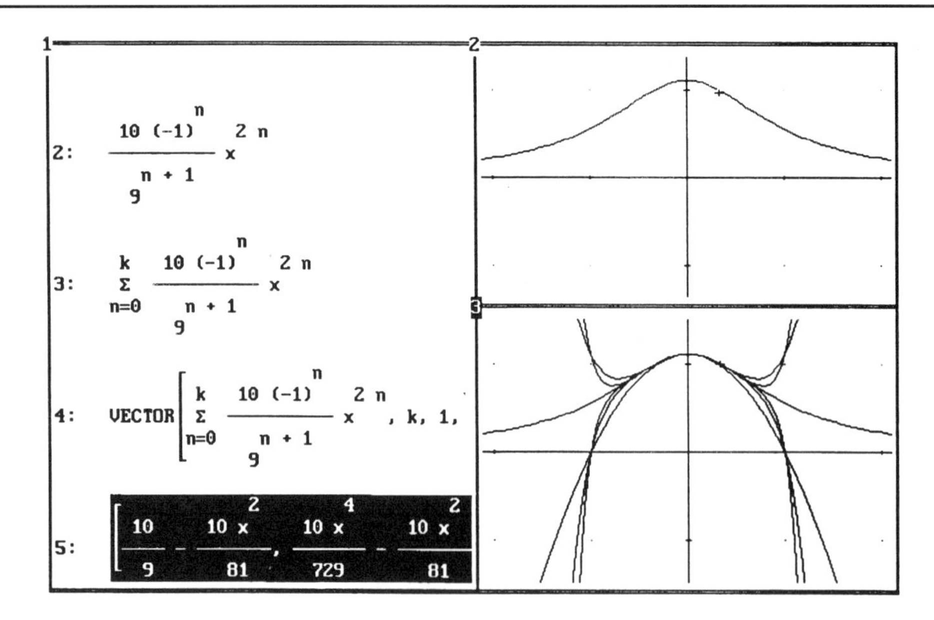

Figure 11.4: **Estimating the radius of convergence, Scale x:3 y:1**

Next **Author** `10(-1)^n/9^(n+1)x^(2n)` and use **Calculus Sum** setting variable:n and Lower limit:0 Upper limit:k. The sum appears in expression 3. We want a list of these sums as k ranges from 1 through 5, so we **Author** `VECTOR(#3, k, 1, 5)` and **Simplify**. The list of sums is partially displayed in expression 5 of Figure 11.4. We **Plot** this to see the picture in window 3 of Figure 11.4. From the picture it appears that the partial sums are close to the function only from about $x = -3$ to $x = 3$, and outside that interval, the partial sums appear to be diverging.

Solution to (b): In order to get the exact value of the radius of convergence, we need to calculate

$$\lim_{n\to\infty} \left| \frac{\frac{10(-1)^{n+1}}{9^{n+2}} x^{2(n+1)}}{\frac{10(-1)^n}{9^{n+1}} x^{2n}} \right|$$

Highlight expression 2 of Figure 11.4 and use **Manage Substitute** to replace n by `n+1`. The result is in expression 6 of Figure 11.5.

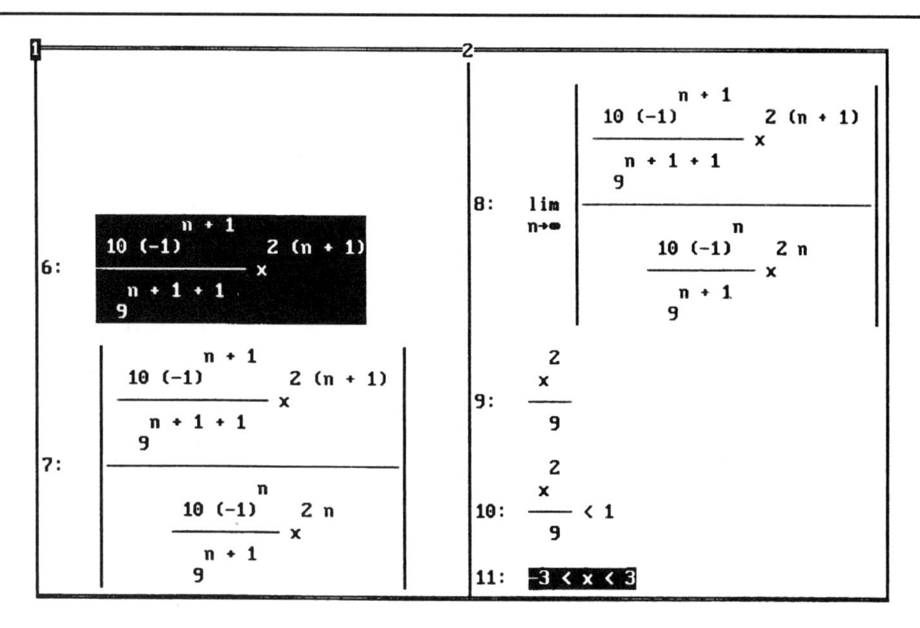

Figure 11.5: **The ratio test for** $\sum_{n=0}^{\infty} \frac{10(-1)^n}{9^{n+1}} x^{2n}$

Now **Author** `ABS(#6/#2)`, and use **Calculus Limit** setting variable:`n` and Point:`inf`. **Simplify** to get the answer, $\frac{x^2}{9}$, in expression 9. The ratio test tells us that the series is sure to converge when this is less than 1. Thus we **Author #9<1** and **soLve**. The solution in expression 11 agrees with our earlier estimate of the radius of convergence. We leave it to the reader to determine what happens at the endpoints.

Solved Problem 11.4: Approximating with Taylor series

New *DERIVE* Lessons •Calculating Taylor polynomials

(a) Using $a = 0$ as the expansion point, find the Taylor polynomial of degree 6 for $\sec x$. Plot the graphs of $\sec x$ and its Taylor polynomial in the same window.

(b) Use the Taylor polynomial you found in (a) to approximate the value of $\sec\left(\frac{1}{2}\right)$. Compare your answer with *DERIVE's* direct approximation of $\sec\left(\frac{1}{2}\right)$.

(c) What is the error predicted by the remainder term for the approximation you made in (b)?

Solution to (a): **Author** `SECx` and **Plot**. The graph is in window 2 of Figure 11.6. To get the Taylor polynomial, use **Calculus Taylor** and press Enter twice to acknowledge expression:#1 and variable:x. When you see the prompt, Degree:5 Point:0, set Degree:`6`. **Simplify** the result to get the Taylor polynomial in expression 3 of Figure 11.6 and then **Plot** to add its graph to the picture. We see that the graph of the Taylor polynomial is close to that of $\sec x$ on the interval $(-\frac{\pi}{2}, \frac{\pi}{2})$, but then it goes off the screen and is not close to the other branches of the graph of $\sec x$. (What does this tell you about the wisdom of using this polynomial to approximate sec(5)?)

Solution to (b): Use **Manage Substitute** to replace x by $\frac{1}{2}$ in the Taylor polynomial. When we **approX**, *DERIVE* returns 1.13934 in expression 5. To check *DERIVE's* answer, we **Author** and **approX** `SEC(1/2)`. From the answer in expression 7 we see that the two answers agree to three places, and that the error in the approximation is 0.00015.

Solution to (b): The maximum error predicted by the remainder term is no more than $\frac{M}{7!}\left(\frac{1}{2}\right)^7$, where M is the maximum of $\left|f^{(7)}(x)\right|$ on the interval $[0, \frac{1}{2}]$. To find M, highlight expression 1 of Figure 11.6 and use **Calculus Differentiate** with Order:7. **Simplify** to get the derivative partially displayed in expression 9 of Figure 11.7.

The graph of the seventh derivative with **Scale** x:0.25 y:1000 appears in window 2 of Figure 11.7. We note that it is increasing on the interval $[0, \frac{1}{2}]$, so that the maximum occurs at the right-hand endpoint, $x = \frac{1}{2}$. Use **Manage Substitute** to replace x by `1/2` in expression 9 and **approX** to get the value 2900.97. In expressions 12 and 13, we have used this value for M in the remainder formula. We see that the error in our approximation in Part (b) is guaranteed to be no more than 0.00449679, but the actual error was less: 0.00015. This illustrates the fact

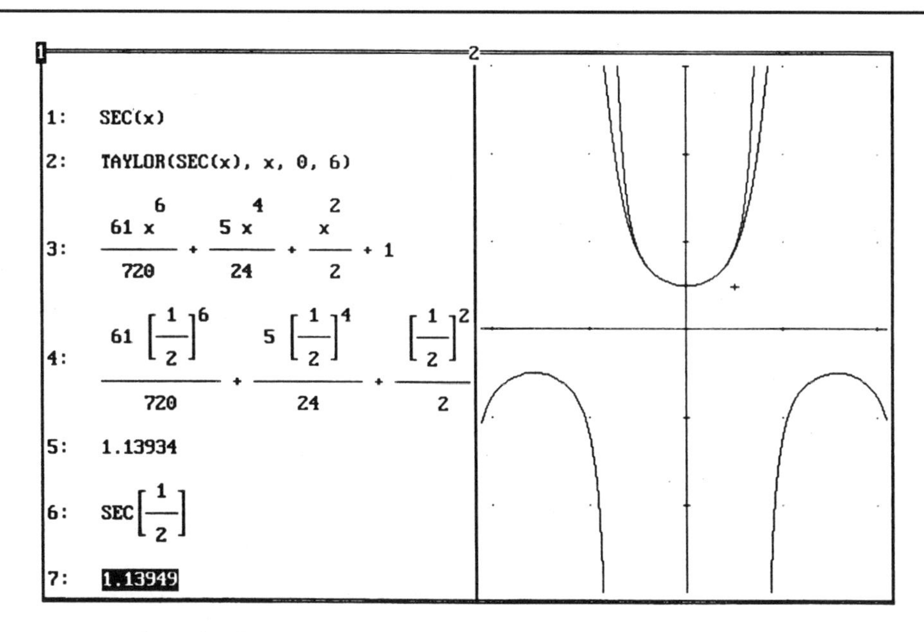

Figure 11.6: **The sixth degree Taylor polynomial for** $\sec x$, **Scale x:2 y:2**

that the remainder term for a Taylor series gives an upper bound for the error but not the exact error.

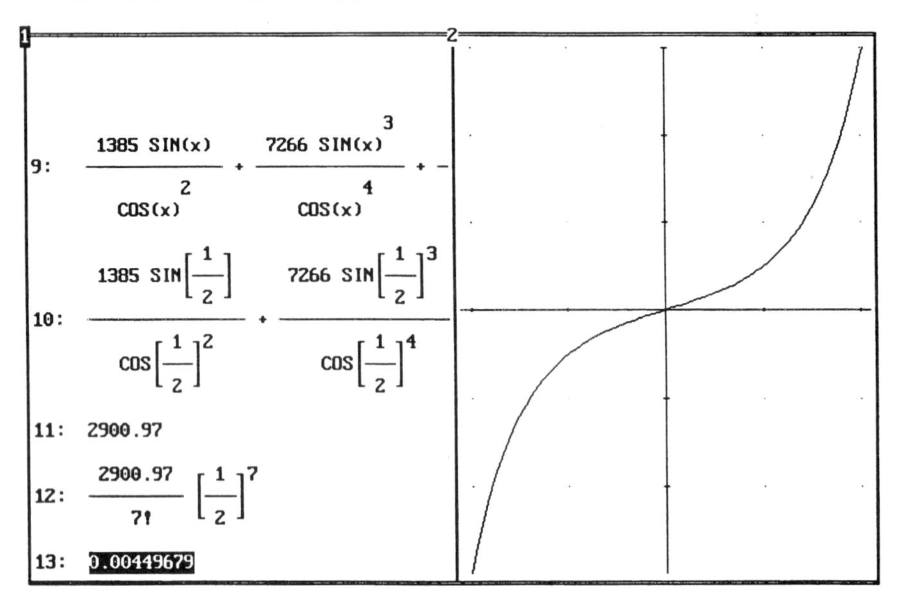

Figure 11.7: **Approximating** sec $\left(\frac{1}{2}\right)$

Practice Problems

1. Find the exact value of $\sum_{n=0}^{\infty} \frac{2^n}{5^n}$. Answer: $\frac{5}{3}$

2. Approximate the value of $\sum_{n=1}^{\infty} \frac{n}{n^3+1}$ by calculating the partial sum with $k = 1000$.
   Answer: 1.11064

3. Plot the graph of the partial sums of $\sum_{n=2}^{\infty} \frac{1}{n + \ln n}$.

4. Plot the graphs of the first five partial sums of $\sum_{n=1}^{\infty} \frac{2^n}{n^2} x^{2n}$.

5. Find the Taylor polynomial about $a = 0$ of degree 5 for $\sin(e^x)$.
   Answer: $-\frac{23x^5}{120} - \frac{25x^4}{120} + \frac{x^2}{2} + x$

6. Find the maximum value of the sixth derivative of $\tan x$ on the interval $[0, 1]$. (Suggestion: We recommend a **Scale** of x:0.5 y:10000 for the graph of the sixth derivative.)
   Answer: 36471.4

Laboratory Exercise 11.1

Approximating Convergent Series

Name ______________________ Due Date ____________

Determine if the following series converge and explain your reasoning. If it converges, approximate the sum. Find the exact value of the sum where possible.

1. $\displaystyle\sum_{n=1}^{\infty} \frac{1}{n(n+1)}$ (Suggestion: Compare with $\displaystyle\sum_{n=1}^{\infty} \frac{1}{n^2}$.)

2. $\displaystyle\sum_{n=1}^{\infty} \frac{1}{n^2 + \ln n}$ (Suggestion: Compare with $\displaystyle\sum_{n=1}^{\infty} \frac{1}{n^2}$.)

3. $\sum_{n=1}^{\infty} \frac{2^n}{n!}$ (Suggestion: Use the ratio test.)

4. $\sum_{n=2}^{\infty} \frac{1}{n \ln n}$ (Suggestion: Use the integral test.)

Laboratory Exercise 11.2

Seeing Convergence Graphically

Name ______________________ Due Date __________

1. Show that $\sum_{n=1}^{\infty} \frac{1}{n^2+n+1}$ converges.

2. Use $\sum_{n=1}^{1000} \frac{1}{n^2+n+1}$ to approximate the sum.

3. Plot the graph of the partial sums of $\sum_{n=1}^{\infty} \frac{1}{n^2+n+1}$. State what variable the horizontal axis corresponds to and what variable the vertical axis corresponds to.

4. Use the graph in Part 3 to estimate $\sum_{n=1}^{\infty} \frac{1}{n^2+n+1}$. Explain how you got your answer and compare it with the answer you obtained in Part 2.

5. It can be shown that the alternating series $\sum_{n=0}^{\infty} \frac{4(-1)^n}{2n+1}$ converges to π.

(a) Use the alternating series test to show that the series converges.

(b) Plot the graph of the partial sums and explain how the graph suggests the convergence of the series. (Suggestion: We recommend a **Scale** of x:3 y:0.1.)

(c) Approximate π by summing the first 1000 terms of the series.

(d) How many terms of the series would you have to sum in order to approximate π correct to six decimal places? (Hint: Recall that the error in using $\sum_{n=0}^{k}(-1)^n a_n$, $a_n > 0$, to approximate $\sum_{n=0}^{\infty}(-1)^n a_n$ is no larger than a_{k+1}.)

Laboratory Exercise 11.3

Euler's Constant

Name ______________________________ Due Date ____________

It is proved in advanced mathematics texts that the sequence $\left\{\sum_{n=1}^{k} \frac{1}{n} - \ln k\right\}$ converges. The number that it converges to is known as Euler's constant, and it is usually denoted by γ.

1. Plot the graph of $\sum_{n=1}^{k} \frac{1}{n} - \ln k$.

2. Explain how this graph suggests the convergence of $\left\{\sum_{n=1}^{k} \frac{1}{n} - \ln k\right\}$, and use the graph to estimate the value of γ.

3. Estimate the value of γ by plugging in several large values of k into $\sum_{n=1}^{k} \frac{1}{n} - \ln k$. Compare your answer with the one you got in Part 2.

Laboratory Exercise 11.4

Radius of Convergence

Name ______________________________ Due Date ____________

It can be shown that $\dfrac{20}{8+x^3} = \sum_{n=0}^{\infty} \dfrac{20(-1)^n}{8^{n+1}} x^{3n}$ when the series converges.

1. On the same screen, plot the graphs of $\dfrac{20}{8+x^3}$ and the first five partial sums of the series.

2. Use the picture in Part 1 to estimate the radius of convergence of the series.

3. Use the ratio test to calculate the radius of convergence exactly and compare your answer with the graphical estimate you found in Part 2.

4. Find the interval of convergence of the series.

Laboratory Exercise 11.5

Approximations Using Taylor Series

Name ______________________________ Due Date __________

1. Find the Taylor polynomial about $a = 0$ of degree 6 for $\tan x$. Plot the graphs of $\tan x$ and the Taylor polynomial on the same screen.

2. If you use the Taylor polynomial from Part 1 to approximate $\tan\left(\frac{1}{2}\right)$, what is the maximum error predicted by the remainder term?

3. Approximate $\tan\left(\frac{1}{2}\right)$ using the Taylor polynomial from Part 1.

4. Use *DERIVE* to directly **approX** the value of $\tan\left(\frac{1}{2}\right)$. What was the exact error in your calculation from Part 3? Is it the same as the error predicted in Part 2? Explain any apparent discrepancies.

5. Let $f(x) = \ln(1 + e^x)$. Plot the graphs of the first 10 derivatives of $f(x)$. (Suggestion: First **Author** `F(x):=LN(1+` Alt e `^x)`. Then **Author** `VECTOR(DIF(F(x),x,k),k,1,10)`. **Simplify** and **Plot**. We recommend a **Scale** of x:1 y:3.)

6. Use the picture you found in Part 5 to find a value of M such that you are sure that $|f^{(n)}(x)| \leq M$ on the interval $[0, 1]$ for every $n \leq 10$. Explain how you got your answer.

7. Using the number M you found in Part 6, calculate the remainder term for $n = 6$ through $n = 9$.

8. Using the expansion point $a = 0$, what degree Taylor polynomial is needed to approximate $\ln(1 + e)$ with error less than 0.0001?

9. Approximate $\ln(1 + e)$ using the Taylor polynomial you found in Part 8. Compare your answer with *DERIVE's* approximation for $\ln(1 + e)$.

Laboratory Exercise 11.6

Machin's Formula

Name ______________________________ Due Date __________

Let $P(x)$ be the Taylor series around $a = 0$ for $4 \arctan x$. It can be shown that the interval of convergence for this series is $(0, 1]$.

1. Show that $P(1) = \pi$ and explain how you would use Taylor polynomials to approximate π.

2. Obtain the Taylor polynomial of degree 29 for $4 \arctan x$, and use it to approximate π. (<u>Note</u>: Recall that *DERIVE's* syntax for $\arctan x$ is `ATANx`.) For use in later parts of this exercise, it will be convenient to give this polynomial a name. Assuming your polynomial is expression 3, **Author** `Q(x):=#3`.)

3. What is the error in the approximation obtained in Part 2?

Before continuing set *DERIVE's* precision using **Options Precision Exact** 50.

A better way to approximate π is by use of *Machin's formula*:

$$\pi = 16 \arctan\left(\frac{1}{5}\right) - 4 \arctan\left(\frac{1}{239}\right)$$

(It's a challenging exercise to use trigonometric identities to prove this formula.)

4. According to Machin's formula, we can approximate π using $4Q\left(\frac{1}{5}\right) - Q\left(\frac{1}{239}\right)$ where $Q(x)$ is the polynomial you defined in Part 2.

5. What is the error in your approximation from Part 4?

6. Explain why Machin's formula provides a better approximation of π than does the method used in Part 2.

Chapter 12

Polar Coordinates and Parametric Equations

New *DERIVE* topics •Plotting polar graphs •Plotting parametric graphs
Calculus concepts •Polar graphs •Parametric graphs •Area •Arc length

Solved Problem 12.1: Polar graphs

New *DERIVE* Lessons •Plotting polar graphs

(a) Plot the graphs of the polar equations $r = \sin^2\theta$ and $r = \cos\theta$.

(b) Find the points of intersection of the graphs.

Solution to (a): We will adhere to the tradition of using θ as the independent variable in polar equations. It can be entered in *DERIVE* using Alt h, but you may use any variable you like. First **Author** `SIN^2`Alt h and then **Plot Beside**. When we get to the plot window, we need to tell *DERIVE* that we want a polar graph. Use **Options State Polar** Enter to do this. Now when we **Plot**, we see the prompt PLOT: Min:-3.1416 Max: 3.1416. *DERIVE* is offering a range for θ to use in plotting the graph. Both $\sin\theta$ and $\cos\theta$ are periodic with period π so if we use the default range, we will see the entire graphs. Press Enter to accept the range. We see the graph in window 2 of Figure 12.1 (it's the "figure eight"). Use **Algebra** to return to your calculations, **Author** `COS`Alt h, and **Plot** once more. We don't have to tell *DERIVE* again that we want a polar graph since the polar option we set earlier remains in effect until we specifically change it. We just **Plot** again and press Enter to accept the default range for θ. We see a graph that appears to be a circle. (Can you prove it is a circle?)

Solution to (b): From the graph in window 2 of Figure 12.1 we see three crossings. One at the origin, one in the fourth quadrant, and another in the first quadrant. To find the latter two we **Author** `#1=#2`. We see from expression 4 that *DERIVE* cannot **soLve** the equation in the exact mode, so we change to the approximate mode using **Options Precision Approximate** and **soLve** once more. To find the zero in the fourth quadrant, we use the range Lower:`-pi/2` Upper:`0`. The solution appears in expression 5. This is the θ value of the intersection. To get the r value, we need to plug this into expression 1 (or expression 2). Highlight expression 1 and use **Manage Substitute** setting variable:`RHS(#5)`. (Recall that `RHS` is the *Right Hand Side* function.) Thus the intersection point is (0.61832, −0.904556). To get the intersection point in

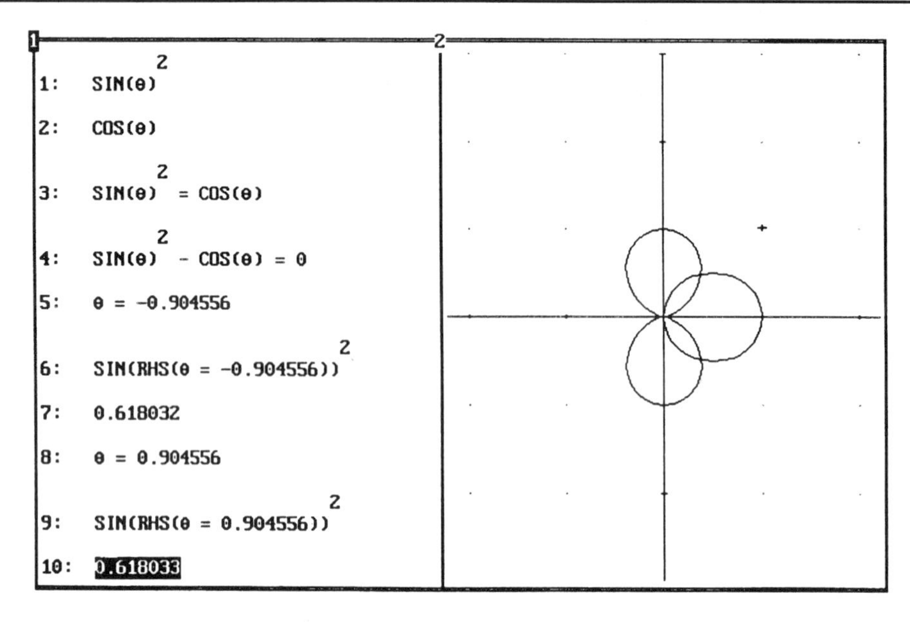

Figure 12.1: **The intersection of polar graphs**

the first quadrant, highlight expression 4 once more and **soLve** on the range Lower:0 Upper: pi/2. From expressions 8 and 10, we see that this intersection point is (0.618033, 0.904556).

The third intersection point $(0, 0)$ cannot be found by equating the right hand sides of the polar equations for the graphs. (Explain why.) Thus it is necessary to plot polar graphs to be sure you have all the intersection points.

Solved Problem 12.2: Area and arc length in polar coordinates

Let R be the region that lies inside the graph of $r = 2 + \sin\theta$.
(a) Find the area of R. (b) Find the arc length of the boundary of R.

Solution to (a): The graph in window 3 of Figure 12.2 shows the region we want. It is important to note that we get the complete graph by letting θ range from 0 to 2π. Recall that in polar coordinates Area $= \dfrac{1}{2}\displaystyle\int_a^b r^2\, d\theta$. Thus we **Author** (1/2)#1^2, and then use **Calculus Integrate** setting Lower limit:0 Upper limit:2pi. We **approX** to get the area in expression 4 of Figure 12.2.

Solution to (b): Recall that in polar coordinates Arc Length $= \displaystyle\int_a^b \sqrt{r^2 + \left(\frac{dr}{d\theta}\right)^2}\, d\theta$. The

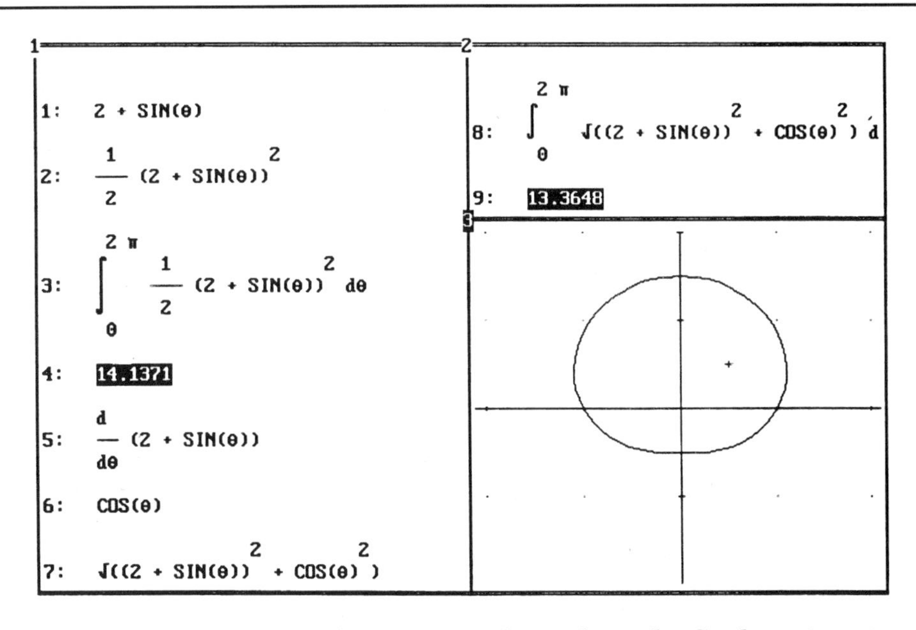

Figure 12.2: **Calculating area and arc length, Scale x:2 y:2**

derivative of expression 1 appears as expression 6. Thus we **Author** `SQRT(#1^2+#6^2)`, and then use **Calculus Integrate** setting Lower limit:`0` Upper limit:`2pi`. We **approX** to get the arc length in expression 9 of Figure 12.2.

Solved Problem 12.3: Arc length for parametric graphs

New *DERIVE* Lessons •Calculating arc length for parametric curves

Plot the graph of the parametric equations $\begin{cases} x = \sin(2t) \\ y = \cos(3t) \end{cases}$, $0 \leq t \leq 2\pi$, and find the arc length of the graph.

Solution: First **Author** `SIN(2t)` and `COS(3t)` as seen in expressions 1 and 2 of Figure 12.3.

To graph the parametric equations $\begin{cases} x = f(t) \\ y = g(t) \end{cases}$, we must enter the functions $f(t)$ and $g(t)$ in a list enclosed by square brackets and separated by commas: $[f(t), g(t)]$. Thus we **Author** `[#1,#2]` as seen in expression 3. **Plot** to open a plot window. If you have been making polar graphs, you need to reset the graphics options to their defaults. Use **Options State**

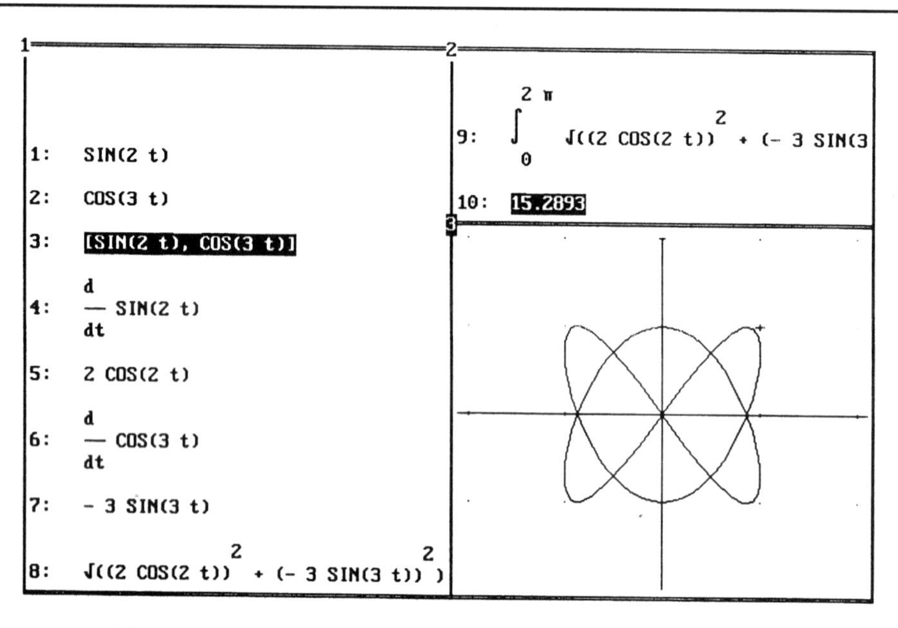

Figure 12.3: **Arc length of a parametric curve**

Rectangular [Enter]. **Plot** again, and as with polar graphs, *DERIVE* offers a range for t, Min:-3.1416 Max:3.1416. Change this to Min:0 Max:`2pi` using the [Tab] key to get to the Max: field. The graph appears in window 3 of Figure 12.3.

To get the arc length, we need the derivatives that appear in expressions 5 and 7. Now **Author** `SQRT(#5^2+#7^2)` as seen in expression 8, and then use **Calculus Integrate** with Lower limit:0 Upper limit:`2pi`. **approX** to get the arc length in expression 10.

Practice Problems

1. Plot the graphs of $r = \sin\theta$ and $r = \cos(2\theta)$ and find the points where they intersect. Answer: $(\frac{1}{2}, -1.08573)$ and $(\frac{1}{2}, 4.22732)$

2. Plot the graph of $\begin{cases} x = \sin t \\ y = \sqrt{t^2+1} \end{cases}$, $0 \le t \le 2$, and find the arc length. Answer: 1.84174

Laboratory Exercise 12.1

Intersections of Polar Graphs

Name ______________________ Due Date ____________

1. Plot the graphs of the polar equations $r = 1 + 2\cos\theta$ and $r = 1 + \cos(2\theta)$ on the same coordinate axes.

2. Find all points where the two graphs intersect. Explain how you got your answer.

3. Find a rectangular equation for $r = 1 + 2\cos\theta$ and solve the equation for y. (Hint: First multiply the polar equation by r. Then use the identities, $r^2 = x^2 + y^2$, and $x = r\cos\theta$. You will need to give *DERIVE* some help to solve for y. Put the square root alone on one side of the equation and square both sides before asking *DERIVE* to **soLve**.)

4. Find all intersection points of the graph of the polar equation $r = 1+2\cos\theta$ with the rectangular graph $y = \cos x$. Express your answers in both rectangular and polar coordinates and explain how you got them.

Laboratory Exercise 12.2

Will Neptune and Pluto Collide?

Name ______________________________ Due Date ____________

If the Sun is located at the origin, then the orbits of Neptune and Pluto are approximated by the polar equations

$$r = N(\theta) = \frac{1.82374 \times 10^{14}}{343\cos(\theta - 0.772727) + 400000}$$

$$r = P(\theta) = \frac{5.52415 \times 10^{13}}{2481\cos(\theta - 3.91232) + 10000}$$

respectively where distances are measured in kilometers.

1. At what point in its orbit is Neptune nearest the Sun? What is the distance at that point? (Hint. You don't need calculus to do this. The fraction is smallest when the denominator is largest.)

2. On the same coordinate axes plot the graphs of the orbits of Neptune and Pluto. (Note: You should think about an appropriate scale before plotting the graphs.)

3. The pictures you made should show two points of potential disaster. Find them and explain how you got your answers.

   Remark: You may guess why the planets do not collide even though the graphs intersect: The orbits of the inner eight planets lie in a plane called *the plane of the ecliptic.* The orbit of Pluto is inclined at an angle of approximately 17 degrees to it.

Laboratory Exercise 12.3

Areas and Arc Length in Polar Coordinates

Name ______________________________ Due Date ____________

Find the area and the arc length of the boundary of each of the regions given below. Explain how you arrive at the limits on the integrals you calculate.

1. The region enclosed by one "leaf" of the "rose" $r = \sin(2\theta)$.

2. The region that lies inside $r = 1 + \cos^2\theta$ and outside the region $r = 1.5 + \sin\theta$.

3. The region that lies inside both $r = 2\sin\theta$ and $r = 1 - \cos\theta$.

4. The region enclosed by the inside loop of $r = \frac{1}{2} + \sin\theta$.

Laboratory Exercise 12.4

Locating Mars in Its Orbit

Name ______________________________ **Due Date** ____________

If the Sun is located at the origin, then the orbit of Mars is approximated by the ellipse

$$r = \frac{1.12495 \times 10^{12}}{476\cos(\theta - 5.85244) + 5000}$$

where distance is measured in kilometers.

1. Plot the graph of the orbit of Mars and calculate the area swept out in one revolution by a "radius" joining the Sun to Mars.

2. The period of Mars is 1.88 years. It is a consequence of *Kepler's second law of planetary motion* that in one year, a radius joining the Sun to Mars will sweep out an area of $\frac{1}{1.88}A$ where A is the area from Part 1. Calculate this area.

3. On January 1, 1960, Mars was located at the polar point corresponding to $\theta = 1.89665$. Where was it on January 1, 1961? (Hint: You need to find θ_1 so that the area swept out from $\theta = 1.89665$ to $\theta = \theta_1$ is the area you found in Part 2.)

4. Plot the graph of the orbit of Mars and locate Mars on the dates from Part 3. Shade in the area swept out during the year.

5. What is the total distance Mars traveled from January 1, 1960 to January 1, 1961?

Laboratory Exercise 12.5

Arc Length for Parametric Functions

Name ______________________________ **Due Date** ____________

1. Plot the graph of $\begin{cases} x = 1 + \cos t \\ y = \sin(2t) \end{cases}$, $0 \leq t \leq 2\pi$ and find the arc length.

2. Find a so that the arc length of $\begin{cases} x = t^2 \\ y = 1 - t^3 \end{cases}$ from $t = 0$ to $t = a$ is 1.

3. A flea attaches itself to the bottom of a 36-inch bicycle wheel. Barbara rides the bike 3 miles down a straight road. The path taken by the flea is known as a *cycloid*, and it is given by the parametric equation $\begin{cases} x = 18(\phi - \sin\phi) \\ y = 18(1 - \cos\phi) \end{cases}$ where ϕ ranges from 0 to $2\pi r$, and r is the total number of revolutions the bicycle wheel makes.

(a) Plot the graph of the path taken by the flea.

(b) How many revolutions did the bicycle wheel make on the 3-mile trip?

(c) Use the arc length formula to determine the total distance the flea traveled.

(d) How far would the flea have traveled if this were an exercise bike that stayed in the living room floor while the wheel turned?

Laboratory Exercise 12.6

The Brachistochrone

Name ______________________________ **Due Date** __________

Begin this exercise by increasing the accuracy with **Options Precision Exact** `15` [Enter].

This is a marble race. Contestants place a marble at the origin and let it roll down a ramp in the form of a path to the point $(\pi, -2)$. Each contestant gets to select the ramp for his or her own marble to follow. The winner is determined by who gets to the bottom in least time.

Harry has chosen the straight line $\begin{cases} x = \pi t \\ y = -2t \end{cases}$, $0 \le t \le 1$, and Judy has chosen the parabolic arc $\begin{cases} x = \pi t \\ y = 2(t-1)^2 - 2 \end{cases}$, $0 \le t \le 1$. The authors of this text have formed a team and chosen the curve $\begin{cases} x = \pi t - \sin(\pi t) \\ y = \cos(\pi t) - 1 \end{cases}$, $0 \le t \le 1$ (it's called a *Brachistochrone*). You may select any curve you want. Just be sure that your parametric curve begins at the origin and goes to the point $(\pi, -2)$ as t ranges from 0 to 1.

1. Which curve do you wish to use? Verify that it starts and ends at the correct places.

2. To decide who wins the race, we need to calculate the time required for a marble to roll down a path $\begin{cases} x = x(t) \\ y = y(t) \end{cases}$, $0 \leq t \leq 1$. Elementary physics can be used to show that this time is

$$\text{Time} = \frac{1}{\sqrt{g}} \int_0^1 \sqrt{\frac{(x'(t))^2 + (y'(t))^2}{-2y(t)}}\, dt$$

where $g = 32$ feet per second is the constant of gravitation near the surface of the Earth.

Find the time required for the marble to roll down each of the four paths and determine who won the race.

3. When you calculated the integrals in Part 2, you may have noticed that *DERIVE* beeped and gave you a "Dubious accuracy" message. Explain why.

4. Based on the outcome of the race, formulate a conjecture about a physical property of the Brachistochrone curve. (This curve and some its properties were studied by Isaac Newton, Johann Bernoulli, and Jakob Bernoulli in the seventeenth century.)

Chapter 13

Vectors and Vector Valued Functions

New *DERIVE* topics	•Calculations with vectors •Solving systems of equations
Calculus concepts	•Vector •Dot product •Cross product •Planes •Tangent vector •Curvature

The symbols $< 1, 2, 3 >$, $(1, 2, 3)$, $\mathbf{i} + 2\mathbf{j} + 3\mathbf{k}$, and $[1, 2, 3]$ are often used to denote the same vector. We will use the latter since that is *DERIVE's* syntax.

Solved Problem 13.1: Calculations with vectors

New *DERIVE* Lessons	•Entering vectors •Calculating length •Calculating dot products •Calculating cross products

Let $\mathbf{u} = [1, 2, 3]$ and $\mathbf{v} = [3, 4, 2]$. Calculate the following.
(a) A unit vector in the direction of $\mathbf{u}$ (b) $\mathbf{u} \cdot \mathbf{v}$ (c) $\mathbf{u} \times \mathbf{v}$

Solution to (a): Each of the items above is easily calculated by hand. The point of this exercise is to introduce *DERIVE's* syntax for these operations. *DERIVE* uses square brackets, `[ ]`, to enclose vectors. Neither `( )` nor `{ }` is acceptable. To define the two vectors, we **Author** `u:=[1,2,3]` and `v:=[3,4,2]` as seen in expressions 1 and 2 of Figure 13.1. To get the unit vector we **Author** and **Simplify** `u/|u|`. The result is in expression 4.

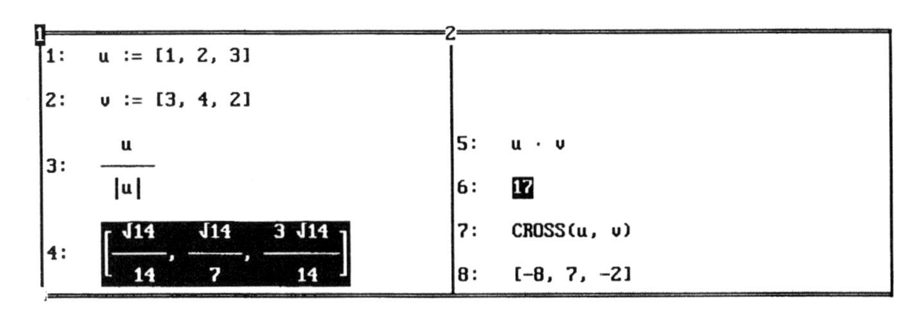

Figure 13.1: **Calculations with vectors**

Solution to (b): To get the dot product, we **Author** `u v` and **Simplify**. (Note: In early versions of *DERIVE*, you must use a period to denote the dot product. If your version of *DERIVE* did not give the correct answer, try `u.v` .)

Solution to (c): To get the cross product, **Author** `CROSS(u,v)` and **Simplify**. The answer is in expression 8 of Figure 13.1.

Solved Problem 13.2: The equation of a plane

New *DERIVE* Lessons •Solving systems of equations

Find the equation of the plane through the points $\mathbf{u} = [3, 1, 1]$, $\mathbf{v} = [1, 4, 2]$, and $\mathbf{w} = [2, 3, 5]$ (a) by solving a system of equations and (b) by finding a vector normal to the plane.

Solution to (a): The equation of the plane is $ax + by + cz = 1$ where a, b, and c are to be determined. Since the plane passes through the three given points, they must satisfy the equation of the plane. Thus we must solve the system of equations

$$\begin{aligned} 3a + b + c &= 1 \\ a + 4b + 2c &= 1 \\ 2a + 3b + 5c &= 1 \end{aligned}$$

First we **Author** each equation on a separate line as seen in expressions 1, 2, and 3 of Figure 13.2. To assemble the system of equations, we **Author** `[#1, #2, #3]`. Now **soLve**, to get

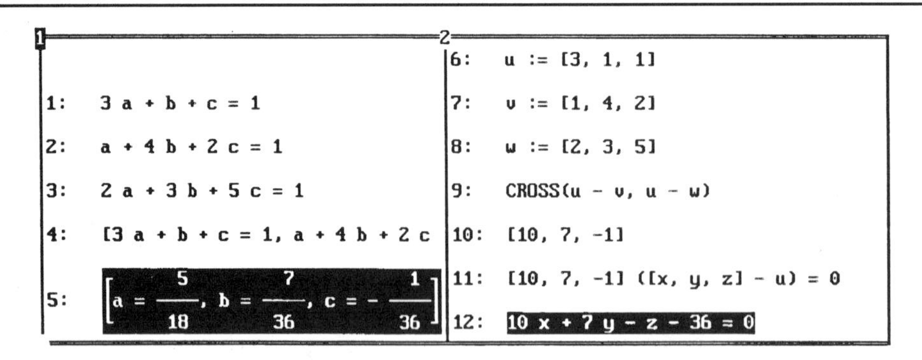

Figure 13.2: **The equation of a plane**

the solution in expression 5 of Figure 13.2. Thus the equation of the plane is $\frac{5}{18}x + \frac{7}{36}y + \frac{1}{36}z = 1$.

Solution to (b): Enter and define the three given vectors as seen in expressions 6, 7, and 8 of Figure 13.2. Since the three points lie on the plane, both the vectors $\mathbf{u} - \mathbf{v}$ and $\mathbf{u} - \mathbf{w}$ are parallel to the plane. Thus the cross product, $(\mathbf{u}-\mathbf{v}) \times (\mathbf{u}-\mathbf{w})$ is normal to the plane. We **Author** and **Simplify** `CROSS(u-v,u-w)`. The result is expression 10. Since the plane passes through the point $\mathbf{u}$, the equation of the plane is $[10, 7, -1] \cdot ([x, y, z] - \mathbf{u}) = 0$. To get this equation, we **Author** and **Simplify** `#10([x,y,z]-u)=0`. The result is in expression 12, and we leave it to the reader to check that this equation is equivalent to the one we found in Part (a).

Notation for space curves

A curve in three dimensional space can be thought of as the range of a function $\mathbf{f}(t)$ defined on a closed interval $[a, b]$. It is also good to think of it as the path of a moving particle which is located at the point $\mathbf{f}(t)$ at time t. There are different notations for this; for example:

$$\mathbf{f}(t) = [\sin t, \cos t, t] \tag{13.1}$$

$$\mathbf{f}(t) = (\sin t)\mathbf{i} + (\cos t)\mathbf{j} + t\mathbf{k} \tag{13.2}$$

$$\begin{aligned} x(t) &= \sin t \\ y(t) &= \cos t \\ z(t) &= t \end{aligned} \tag{13.3}$$

all represent the same curve (a helix). We will use (13.1) because it is *DERIVE's* syntax, but you should be comfortable with all of them.

Solved Problem 13.3: Arc length of space curves

New *DERIVE* Lessons •Calculating the derivative of a vector valued function

(a) Find the length of the arc of the curve defined by $\mathbf{f}(t) = [t, t^2, t^3]$ from $t = 0$ to $t = 2$.

(b) Find a so that the arc length from $t = 0$ to $t = a$ is 1.

Solution to (a): The arc length formula for a space curve $\mathbf{f}(t) = [x(t), y(t), z(t)]$ is

$$\int_a^b |\mathbf{f}'(t)|\, dt = \int_a^b \sqrt{(x'(t))^2 + (y'(t))^2 + (z'(t))^2}\, dt$$

The quantity $|\mathbf{f}'(t)|$ represents the *speed* of our moving particle, so the integral gives the distance it travels, which is the arc length. We **Author** `[t, t^2, t^3]` as seen in expression 1 of Figure 13.3. To get the derivative, we use **Calculus Differentiate** as usual. The derivative

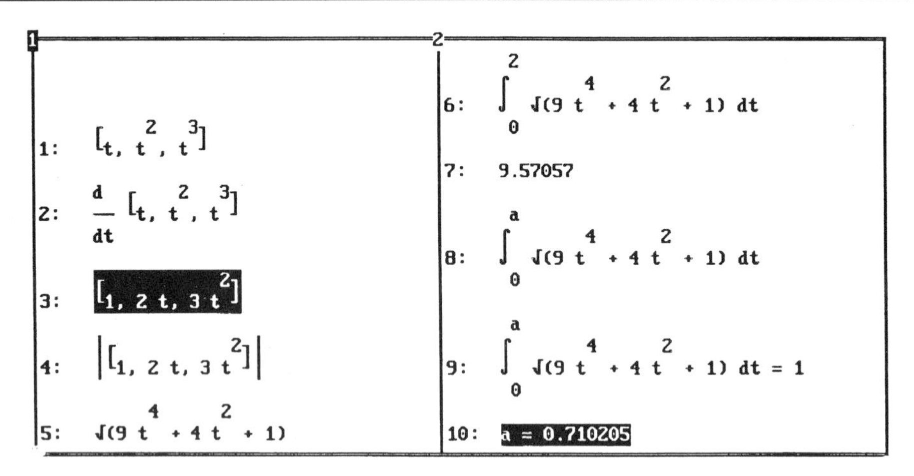

Figure 13.3: **Calculating arc length**

appears in expression 3. To get the speed we **Author** `|#3|` and **Simplify**. The result is in expression 5. We integrate it to get the arc length in expression 7.

Solution to (b): To get the arc length from $t = 0$ to $t = a$, we highlight expression 5 of Figure 13.3 and use **Calculus Integrate** with Lower limit:0 Upper limit:a. We want this integral to be 1, so we **Author** `#8=1`. *DERIVE* cannot solve the equation in expression 9 in the exact mode, so we change to the approximate mode using **Options Precision Approximate**. Now, we know that the value of a is somewhere between 0 and 2 (explain why), so we **soLve** on the range Lower:0 Upper:2. The solution is in expression 10 of Figure 13.3.

Solved Problem 13.4: Invariants of curves

Let $\mathbf{f}(t) = [t^3 + 1, t^2 - 1]$. Calculate the following at $t = 2$.
(a) The unit tangent vector (b) The unit normal vector (c) The curvature

Solution to (a): We **Author** `[t^3+1, t^2-1]` (not shown in Figure 13.4) and use **Calculus Differentiate** to get the tangent vector in expression 3. To get the unit tangent, we **Author** `#3/|#3|` and **Simplify**. The result is in expression 5. Now we use **Manage Substitute** to replace t by 2 and **approX** to get the answer in expression 7.

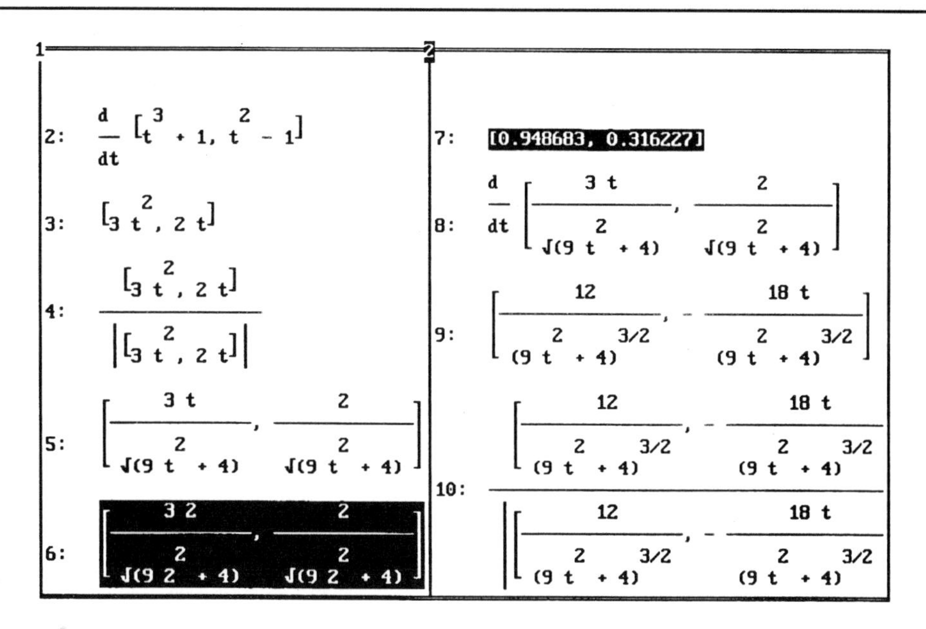

Figure 13.4: **The unit tangent vector**

Solution to (b): To get the unit normal, we first need to differentiate the unit tangent. Highlight expression 5 of Figure 13.4, use **Calculus Differentiate**, and the derivative appears in expression 9. We **Author** `#9/|#9|` and **Simplify** to get the unit normal in expression 11 of Figure 13.5. When we put in $t = 2$ we get the answer in expression 13. You should verify that the unit tangent in expression 7 of Figure 13.4 is in fact orthogonal to the unit normal in expression 13 of Figure 13.5.

Solution to (c): The curvature of $\mathbf{f}$ at t is $\dfrac{|\mathbf{T}'(t)|}{|\mathbf{f}'(t)|}$ where $\mathbf{T}$ is the unit tangent. $\mathbf{T}'(t)$ appears in expression 9 of Figure 13.4, and $\mathbf{f}'(t)$ is in expression 3 of Figure 13.4. Thus we **Author** `|#9|/|#3|` to get the curvature in expression 15 of Figure 13.5. When we plug in $t = 2$ to get the curvature at 2, *DERIVE* returns 0.0118585 in expression 17.

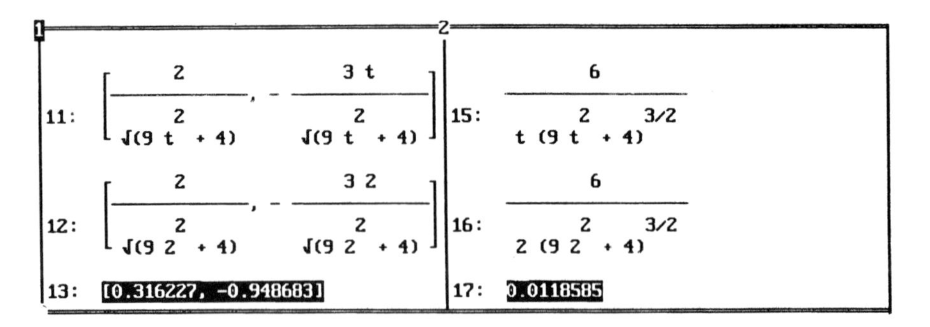

Figure 13.5: **The unit normal vector and the curvature**

Practice Problems

1. Let $\mathbf{u} = [3, 4, 2]$ and $\mathbf{v} = [2, 5, 5]$. Use *DERIVE* to calculate the following.

   $3\mathbf{u} - 2\mathbf{v}$ Answer: $[5, 2, -4]$ $\qquad$ $\mathbf{u} \cdot \mathbf{v}$ Answer: 36 $\qquad$ $\mathbf{u} \times \mathbf{v}$ Answer: $[10, -11, 7]$

2. Solve the system of equations

$$\begin{aligned} x + y + z &= 3 \\ 2x - y + 4z &= 8 \\ 3x + 4y - z &= 4 \end{aligned}$$

   Answer: $x = \dfrac{3}{2}$, $y = \dfrac{1}{5}$, $z = \dfrac{10}{13}$

3. For $\mathbf{f}(t) = [1, t, t^2]$, calculate the following.

   (a) The arc length from $t = 0$ to $t = 1$ Answer: 1.47894

   (b) The unit tangent at $t = 1$ Answer: $[0, 0.447213, 0.894427]$

   (c) The unit normal at $t = 1$ Answer: $[0, -0.894427, 0.447213]$

   (d) The curvature at $t = 1$ Answer: 0.178885

Laboratory Exercise 13.1

Equations of Planes

Name ________________________________ Due Date ____________

1. Find the equation of the plane passing through the points $[3,4,1]$, $[2,5,5]$, and $[7,2,4]$ (a) by solving a system of equations, and (b) by finding a normal vector to the plane. Show that the two equations are equivalent.

2. Find the distance from the plane in Part 1 to the point $[3,8,9]$.

3. Find all planes that pass through the points $[1, 1, 2]$ and $[2, 0, 1]$ and are a distance $\frac{1}{2}$ from the point $[1, 2, 3]$.

Laboratory Exercise 13.2

Changing Parameters

Name ______________________________ Due Date ____________

Let $\mathbf{f}(t) = [t^2, t^3]$ and $\mathbf{g}(t) = [(5t)^2, (5t)^3]$.

1. How do the curves defined by $\mathbf{f}$ and $\mathbf{g}$ for $0 \le t \le \infty$ compare? (*DERIVE* will plot them for you.)

2. Assuming $\mathbf{f}(t)$ and $\mathbf{g}(t)$ each represent the location of moving particles at time t, how do their speeds compare?

3. We are going to introduce a new parameterization for $\mathbf{f}$ using arc length as the parameter. The steps are as follows.

   (a) Use **Declare Variable a Positive**. This will tell *DERIVE* that a is positive. Now, calculate the arc length of $\mathbf{f}$ from $t = 0$ to $t = a$.

(b) Set the answer you got in Part 3(a) equal to s and solve the resulting equation for a. (Note: You should get two solutions; be sure to use the positive one.)

(c) If $a(s)$ denotes the solution in Part 3(b), define $\mathbf{h}(s) = \mathbf{f}(a(s))$.

4. Use *DERIVE* to plot $\mathbf{h}(\mathbf{t})$. How does it compare with $\mathbf{f}(\mathbf{t})$?

5. Verify that the speed of $\mathbf{h}$ is always 1.

Laboratory Exercise 13.3

Space Curve Invariants

Name ____________________ Due Date __________

Let $\mathbf{f}(t) = [t, \sin t, \sin(2t)]$

1. Calculate the unit tangent vector for $\mathbf{f}$ at $t = 3$.

2. Calculate the unit normal vector for $\mathbf{f}$ at $t = 3$.

3. Verify that the vectors you found in Parts 1 and 2 are perpendicular.

4. Calculate the curvature of $\mathbf{f}$ at $t = 3$.

Chapter 14

Partial Derivatives

New *DERIVE* topics •Plotting 3-dimensional graphs •Calculating limits •Calculating partial derivatives •Calculating the gradient
Calculus concepts •Functions of several variables •Limits •Partial derivatives •Gradient •Directional derivatives •Tangent plane •Maxima and Minima

Solved Problem 14.1: Graphs and level curves

New *DERIVE* Lessons •Plotting 3-dimensional graphs

Let $f(x,y) = \frac{y^3}{x^2+1}$. (a) Plot the graph of f and (b) Plot several level curves of f.

Solution to (a): **Author** `y^3/(x^2+1)` and **Plot Beside**. *DERIVE* will recognize that we are dealing with a 3-dimensional graph and open a 3-D plot window. **Plot** again to see the graph. In window 2 of Figure 14.1 we have set the **Grids** to x:30 y:30 to make a better picture. (Use the Tab key to set the y: field.)

Solution to (b): Before proceeding, we should close the 3-D plot window. Otherwise, when we ask *DERIVE* to plot the level curves, it will open a third window. From the plot window, use **Window Close** to do that. To get the level curves, it is tempting to **Plot** $\frac{y^3}{x^2+1} = c$ for different values of c, but *DERIVE* will beep and say "Cannot do implicit plots." Therefore we must try something else: We **Author** `#1=c` and **soLve** for **y**. The solution appears in expression 3 of Figure 14.2. *DERIVE* will plot this if we plug in specific values of c. To make a list of these curves for several choices of c, we **Author** `VECTOR(#3,c,-3,3,0.5)`. Before we **Simplify**, we use **Manage Branch Real** to change the way *DERIVE* handles cube roots. If we ask *DERIVE* to **Simplify** $(-3)^{\frac{1}{3}}$ without doing this it will return the complex number $\frac{3^{\frac{1}{3}} + 3^{\frac{5}{6}}i}{2}$ rather than $-3^{\frac{1}{3}}$. Now we **Simplify** expression 6 and **Plot** the result to see the level curves in window 2 of Figure 14.2.

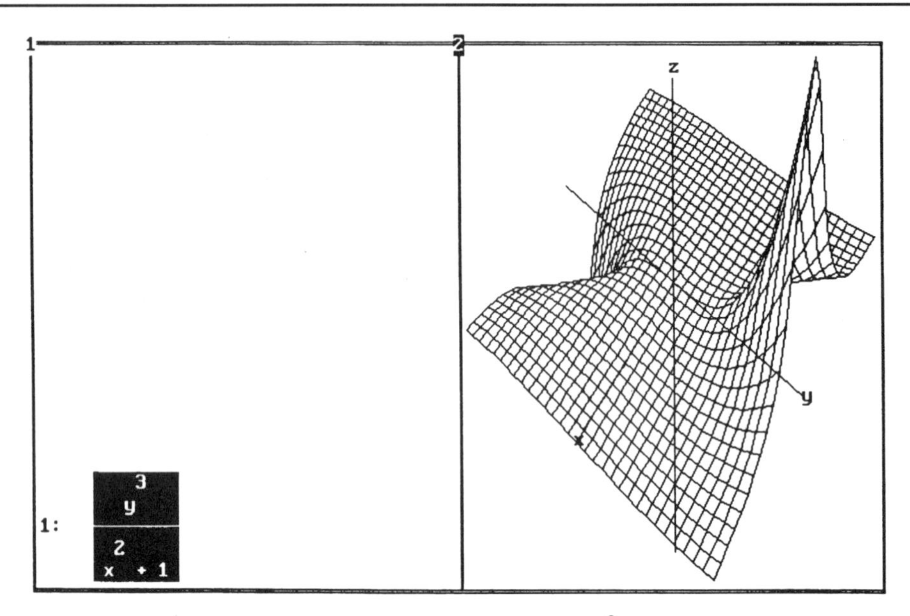

Figure 14.1: **Graph of** $\dfrac{y^3}{x^2+1}$

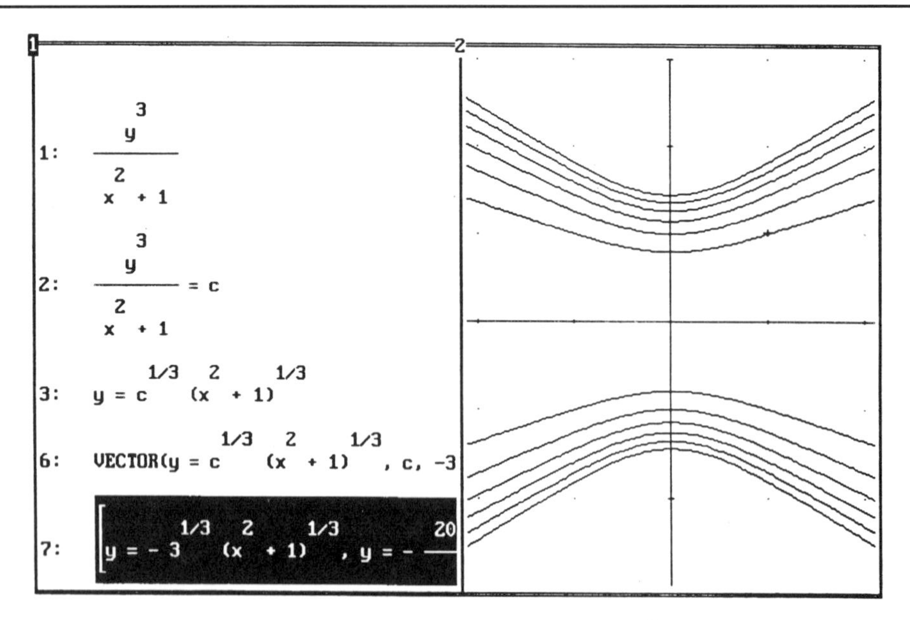

Figure 14.2: **Level curves of** $\dfrac{y^3}{x^2+1}$

Solved Problem 14.2: Limits

New *DERIVE* Lessons •Calculating limits of functions of several variables

Calculate the following limits if they exist.

(a) $\lim_{(x,y)\to(0,0)} \frac{\sin(x^2+y^2)}{x^2+y^2}$ (b) $\lim_{(x,y)\to(0,0)} \frac{xy}{2x^2+3y^2}$

Solution to (a): **Author** `SIN(x^2+y^2)/(x^2+y^2)` as seen in expression 1 of Figure 14.3. We cannot get two-variable limits from the menu as we can with single-variable limits, so we **Author** `LIM(#1, [x,y], [0,0])`. When we **Simplify**, *DERIVE* returns the value, 1, in expression 3.

When we ask *DERIVE* to calculate $\lim_{[x,y]\to[a,b]} f(x,y)$, it actually calculates $\lim_{y\to b}\left(\lim_{x\to a} f(x,y)\right)$. This is correct provided the two-variable limit exists, but may be incorrect otherwise. Thus we need to verify that the two-variable limit exists. In window 3 of Figure 14.3 we have plotted the graph of $\frac{\sin(x^2+y^2)}{x^2+y^2}$ setting **Grids** to x:40 y:40. The graph suggests that the limit does indeed exist. One way to verify that it really does in this case is to change to polar coordinates. Since

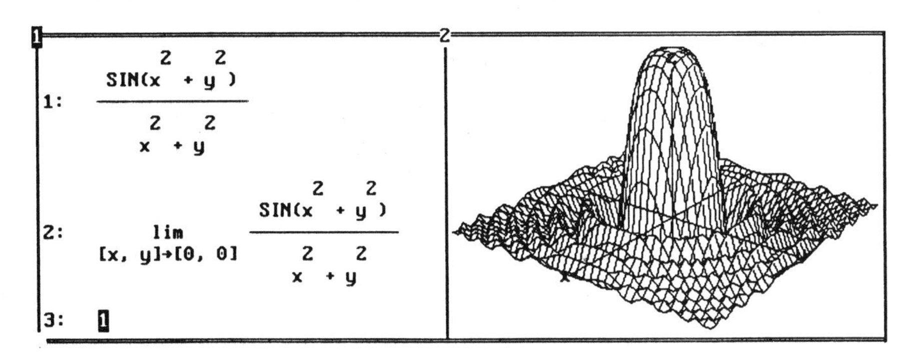

Figure 14.3: **The graph of** $\frac{\sin(x^2+y^2)}{x^2+y^2}$

$x^2+y^2=r^2$ we get $\frac{\sin(r^2)}{r^2}$. Letting (x,y) go to $(0,0)$ is the same as letting r go to 0 and we know that $\lim_{r\to 0}\frac{\sin(r^2)}{r^2}$ exists and equals 1. (We did not use the computer for this, but *DERIVE* can correctly calculate this one-variable limit.)

Solution to (b): We **Author** `xy/(2x^2+3y^2)` as expression 1 and then **Author** `LIM(#1,[x,y],[0,`
When we **Simplify**, *DERIVE* returns the value 0 in expression 3 of Figure 14.4. We warned in (a) that we must check to see if two-variable limits exist, and in window 2 we have plotted the graph setting **Grids** to x:40 y:40 to check this. The graph makes us doubt the answer. In fact, it appears that we can get different limits by approaching the origin along different lines.

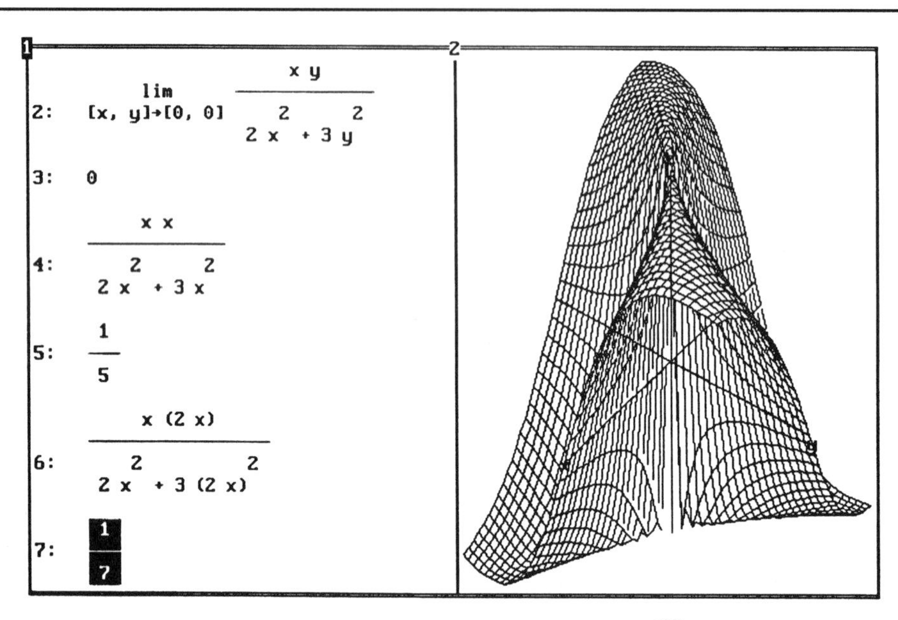

Figure 14.4: **Calculating** $\lim_{(x,y)\to(0,0)} \frac{xy}{2x^2+3y^2}$

To check the limit along the line $y = x$, we highlight the original function and use **Manage Substitute** putting **x** in place of y. From expression 5, we see that the function has the constant value of $\frac{1}{5}$ along the line $y = x$, and so the limit along this line is $\frac{1}{5}$.

We repeat this procedure putting **2x** in place of y to check the limit along the line $y = 2x$. We see from expression 7 that this limit is $\frac{1}{7}$. Since we get different limits along different lines, we conclude that the limit does not exist and that *DERIVE's* answer in expression 3 is incorrect.

We reiterate that *DERIVE* gives the correct answer to $\lim_{[x,y]\to[a,b]} f(x,y)$ *if the limit exists,* but may give an incorrect answer otherwise. Thus we must be sure the limit exists before we calculate it.

Solved Problem 14.3: Partial derivatives and the gradient

New *DERIVE* Lessons •Calculating partial derivatives •Calculating the gradient

Let $f(x, y) = \sin(xy)$. (a) Verify that $f_{xy} = f_{yx}$. (b) Calculate ∇f (the gradient of f), and find the equation of the tangent plane to the graph of f where $x = 1$ and $y = 2$.
(c) Calculate the directional derivative, $D_{\mathbf{u}}f(1, 2)$ where $\mathbf{u}$ is a unit vector in the direction of $(3, 4)$.

Solution to (a): **Author** `SIN(xy)` as seen in expression 1 of Figure 14.5. To get f_x, use

```
1:  SIN(x y)

    d
2:  — SIN(x y)
    dx

3:  y COS(x y)

    d
4:  — (y COS(x y))
    dy

5:  COS(x y) - x y SIN(x y)

    d
6:  — SIN(x y)
    dy

7:  x COS(x y)

    d
8:  — (x COS(x y))
    dx

9:  COS(x y) - x y SIN(x y)

10: GRAD(SIN(x y), [x, y])

11: [y COS(x y), x COS(x y)]

12: SIN(1 2)

13: 0.909297

14: [2 COS(1 2), 1 COS(1 2)]

15: [-0.832293, -0.416146]

16: z = [-0.832293, -0.416146] ([x, y] - [1, 2]

             8670 x     4335 y     165243288
17: z = -  ------- - ------- + ---------
             10417      10417      64199971

18: z = - 0.832293 x - 0.416146 y + 2.57388
```

Figure 14.5: **Partial derivatives and the gradient**

Calculus Differentiate with variable: x, and **Simplify**. The result is in expression 3. As we see, *DERIVE's* **Differentiate** command yields partial derivatives.

To get f_{xy}, we need the partial derivative of f_x with respect to y. Thus we use **Calculus Differentiate** with variable:y. The result is in expression 5. In expressions 6 through 9, we have reversed the order of differentiation, and we note that expressions 5 and 9 of Figure 14.5 are identical.

Solution to (b): Since f_x is in expression 3 of Figure 14.5 and f_y is in expression 7, we can use `[#3, #7]` to make the gradient. *DERIVE* also has an internally defined GRAD function that calculates the gradient directly. We **Author** `GRAD(#1,[x,y])` and **Simplify**. The result is in expression 11.

There are several equivalent forms for the equation of the tangent plane. We will use $z = \nabla f(x_0, y_0) \cdot ([x, y] - [x_0, y_0]) + f(x_0, y_0)$. We are using $x_0 = 1$ and $y_0 = 2$, so we need to evaluate the function and the gradient at these points. Use **Manage Substitute** with expressions 1 and 11 of Figure 14.5 to do this. $f(1, 2)$ appears in expression 13, and $\nabla f(1, 2)$ appears in expression 15. Thus we **Author** `z=#15.([x,y]-[1,2])+#13`. **Expand** and then **approX** to get the equation of the tangent plane in expression 18 of Figure 14.5.

Solution to (c): A unit vector in the direction of $[3, 4]$ is calculated in expressions 19 through 21 of Figure 14.6. Thus the directional derivative is given by $\nabla f(1, 2) \cdot [0.6, 0.8]$. We **Author** and **approX** `#15.#21`, and the directional derivative appears in expression 23.

```
19:  [3, 4]

      [3, 4]
20:  --------
     |[3, 4]|

21:  [0.6, 0.8]

22:  [-0.832293, -0.416146] [0.6, 0.8]

23:  -0.832293
```

Figure 14.6: **Calculating a directional derivative**

Solved Problem 14.4: Finding maxima and minima

Plot the graph of $x^3 + y^2 - xy - x$, find the critical points, and determine if each is a local maximum, local minimum, or saddle point.

Solution: **Author** `x^3+y^2-xy-x` and **Plot** setting **Grids** to x:30 y:30. The graph appears in window 2 of Figure 14.7. The partial derivatives appear in expressions 3 and 5.

The critical points occur where expressions 3 and 5 are both zero.

$$3x^2 - y - 1 = 0 \qquad (14.1)$$

$$2y - x = 0 \qquad (14.2)$$

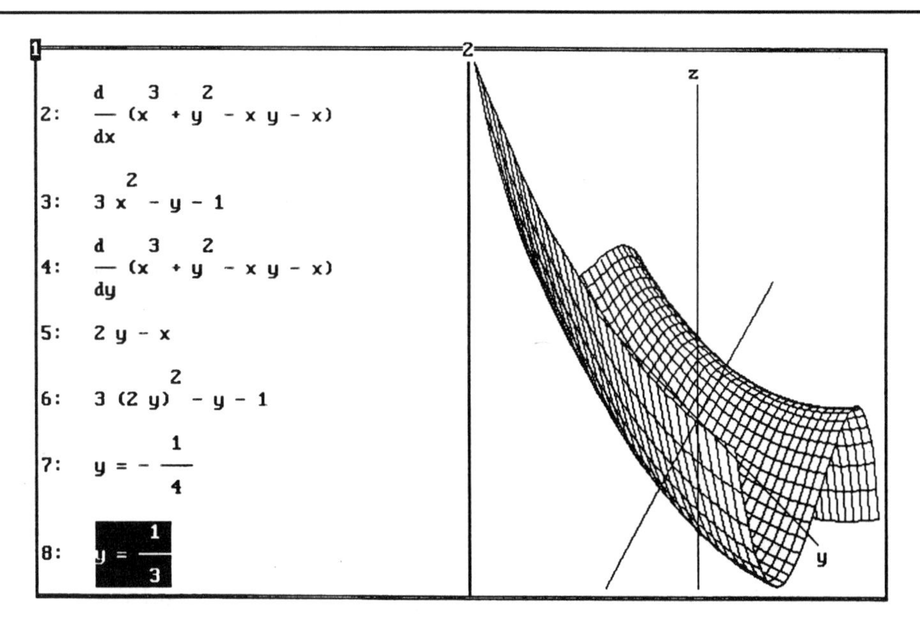

Figure 14.7: **Finding the critical points of $x^3 + y^2 - xy - x$**

From (14.2), we see that $x = 2y$. Thus we highlight expression 3 and use **Manage Substitute** to replace x by 2y. The result is in expression 6 which we **soLve** to get the y-coordinates of the critical points in expressions 7 and 8. Since $x = 2y$, we conclude that the critical points are $\left(-\frac{1}{2}, -\frac{1}{4}\right)$ and $\left(\frac{2}{3}, \frac{1}{3}\right)$.

To test a critical point, we need to evaluate $f_{xx}(x, y)$ and the *discriminant function*, $D(x, y)$, at the critical points.

$$D(x, y) = f_{xx}(x, y) f_{yy}(x, y) - f_{xy}^2(x, y)$$

If you plan to do several such problems, you can save yourself a lot of work by writing some code. Begin with a clear screen and **Author** the following.

```
F(x,y):=

DIF(F(x,y),x,2)

DIF(F(x,y),y,2)

DIF(DIF(F(x,y),x),y)
```

```
[#2,#2 #3 - #4^2]

LIM(#5,[x,y],[a,b])

TEST(a,b):=#6
```

You may wish to save this as a file using **Transfer Save Derive** TEST. You can retrieve it when you need it using **Transfer Merge** TEST.

We **Author** `F(x,y):=x^3+y^2-xy-x` to tell *DERIVE* which function we want. To test the critical point $\left(-\frac{1}{2},-\frac{1}{4}\right)$, we **Author** and **approX** `TEST(-1/2, -1/4)`. The first coordinate of the answer in expression 13 of Figure 14.8 is $f_{xx}\left(-\frac{1}{2},-\frac{1}{4}\right)$ and the second coordinate is the discriminant function evaluated at the critical point. Since the second coordinate is negative, we conclude that this critical point corresponds to a saddle point.

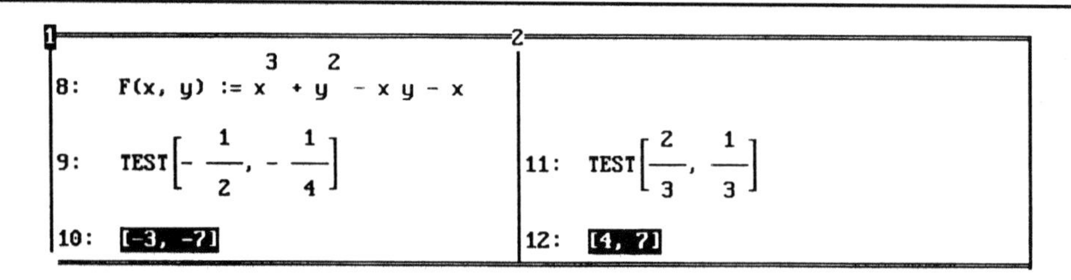

Figure 14.8: **Testing the critical points**

To test $\left(\frac{2}{3},\frac{1}{3}\right)$, we **Author** and **approX** `TEST(2/3, 1/3)`. We see from expression 15 of Figure 14.8 that both f_{xx} and the discriminant function are positive. We conclude that this critical point corresponds to a minimum. These conclusions are verified by the graph in window 2 of Figure 14.7.

Solved Problem 14.5: A box of least cost

A decorative box is to have a wooden base costing $3 per square foot, two opposite gold-plated sides costing $39 per square foot, two opposite silver-plated sides that cost $23 per square foot, and a platinum top that costs $61 per square foot. The volume of the box is to be 1 cubic foot. Find the dimensions and the cost of the most economical box.

Solution: We will use h for the height of the box, s for the length of the base of the silver plate, and g for the length of the base of the gold plate. In terms of these variables, the area of each plate is

Wood	gs
Silver	sh
Gold	gh
Platinum	gs

Thus, since there are two silver and two gold plates, the cost of the box is $3sg + 2(23sh) + 2(39gh) + 61sg$. Now the volume, sgh, of the box is 1 cubic foot so that $h = \frac{1}{sg}$. This gives the cost as a function of two variables $64sg + \frac{46}{g} + \frac{78}{s}$.

Author the cost function as **64sg+46/g+78/s**. The partial derivatives appear in expressions 3 and 5 of Figure 14.9. We **soLved** expression 3 for **s** to get expression 6 and then plugged

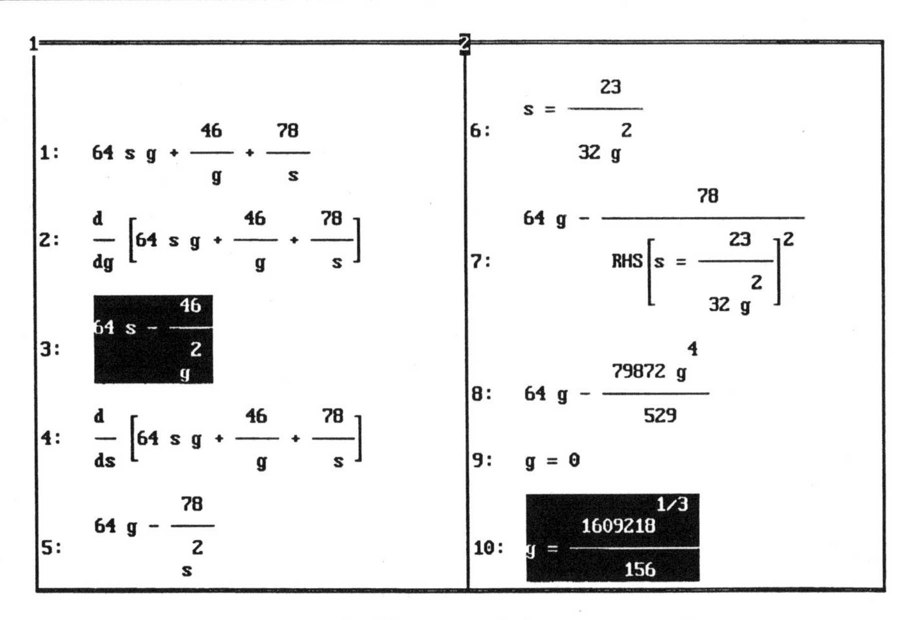

Figure 14.9: **Finding a minimum cost box**

this value for s into expression 5 to get expression 7. When we **soLve** expression 7, we get expressions 9 and 10 and two additional complex solutions that are not shown in Figure 14.9 and that we ignore. We discard the solution $g = 0$ because our cost function is not defined there. When we **approX** expression 10, we get $g = 0.751185$, and this gives $s = 1.27374$ and $h = 1.04512$, yielding a cost of \$183.71.

You may, if you wish, test this point using the TEST.MTH file to be sure that it gives a minimum, but that is really not necessary. We know from the physical constraints on the problem that there is a minimum cost, and we have only one critical point. We conclude that it must give the minimum cost.

Practice Problems

1. Plot the graph of $\dfrac{xy}{1+x^2+y^2}$.

2. Use *DERIVE* to calculate $\lim_{(x,y)\to(1,2)} x^2 + y^2$. Answer: 5

3. Let $f(x,y) = e^{xy}$. Use *DERIVE* to calculate f_{xxy}, the gradient of f, and $D_{[6,8]}f(0,1)$.
   Answer: $f_{xxy} = e^{xy}(xy^2 + 2y)$, $\nabla f = [ye^{xy}, xe^{xy}]$, and $D_{[6,8]}f(0,1) = 0.6$.

Laboratory Exercise 14.1

Graphs and Level Curves

Name ______________________________ Due Date ____________

Plot the graph of each of the following functions and plot several level curves.

1. $f(x, y) = \dfrac{x^2}{x^2 + y^2 + 1}$

2. $f(x, y) = \dfrac{x}{e^y}$

3. $f(x,y) = \dfrac{x^2 + y^2}{y^4 + 1}$

4. $f(x,y) = \sin(x^2 + y^2)$

Laboratory Exercise 14.2

Limits of Functions of Two Variables

Name ______________________ Due Date ____________

For each of the following functions

(a) Use *DERIVE* to calculate the limit as (x, y) goes to $(0, 0)$.

(b) Plot the graph and explain how it does or does not support the calculation in Part (a).

(c) Determine if the answer from Part (a) is correct and justify your answer.

1. $\sqrt{x^2 + y^2}\ln(x^2 + y^2)$ (Suggestion for Part (c): Change to polar coordinates.)

2. $\dfrac{x}{x + y}$ (Suggestion for Part (c): Check along the lines $y = x$ and $y = 2x$.)

3. $\dfrac{xy}{\ln(x^2+y^2+1)}$ (Suggestion for Part (c): Check along the line $y = x$.)

4. $(1+x^2+y^2)^{\frac{1}{x^2+y^2}}$ (Suggestion for Part (c): Change to polar coordinates.)

Laboratory Exercise 14.3

Applications of the Gradient

Name ______________________________ **Due Date** __________

Let $f(x, y) = e^{xy} \ln(x + y)$ be the temperature of the xy-plane at the point (x, y) in degrees Celsius. Assume distance is measured in meters.

1. What is the temperature at (2, 1)?

2. Calculate ∇f.

3. If I am standing at (2, 1), in what direction should I move to warm up the fastest? What direction will make me cool off the fastest?

4. Find all unit vectors **u** such that if I move in the direction of **u** my temperature will go up at 3 degrees per meter.

5. In what direction should I move so that my temperature will not change. (Hint: Recall that the gradient vector is orthogonal to the level curves.)

Laboratory Exercise 14.4

The Heat Equation

Name ______________________________ Due Date __________

Consider the x-axis as a long metal rod that has an initial temperature given by a function $f(x)$. If the rod is left to cool, it can be shown that the function $H(x,t)$, which gives the temperature of the rod at time t, satisfies a partial differential equation known as the *heat equation.*

$$\frac{\partial^2 H}{\partial x^2} = \frac{\partial H}{\partial t}$$

1. Show that $G(x,t) = \dfrac{e^{-\frac{x^2}{4t}}}{2\sqrt{\pi t}}$ is a solution of the heat equation.

2. It can be shown that with mild restrictions on f, $F(x,t) = \int_{-\infty}^{\infty} G(x-u,t)f(u)\,du$ is a solution of the heat equation that satisfies the initial condition $F(x,0) = f(x)$. Find a solution of the heat equation that satisfies the initial condition $F(x,0) = x^4$. Plot the graph of the function you get, verify that it is a solution of the heat equation, and show that it satisfies the initial condition.

3. Suppose the initial temperature of the rod is 0 except on the interval $[-1,1]$, where it is 1. Find a function that gives the temperature of the rod at time t. Verify that it is a solution of the heat equation and plot its graph. (Note: When you integrate, *DERIVE* will give an answer in terms of the ERF function which is defined by $\text{ERF}(x) = \frac{2}{\sqrt{\pi}} \int_0^x e^{-t^2}\,dt$.)

Laboratory Exercise 14.5

Maxima and Minima of Functions of Two Variables

Name ______________________________ **Due Date** ____________

For each of the following functions, (a) plot the graph, (b) find the critical points, and (c) test each critical point to determine if it is a maximum, minimum, or saddle point.

1. $4xy - x^3 - y^3$

2. $x^3 + y\cos x - xy$

3. $xy + \ln(x^2 + y^2 + 1)$

4. $\dfrac{x + y}{x^2 + 2y^2 + 1}$

Laboratory Exercise 14.6

Applications of Extrema of Two-Variable Functions

Name ______________________________ **Due Date** __________

1. **A metal box**: A metal box with no lid is to be assembled from rectangular pieces of metal and is to have a volume of 8 cubic feet. The base of the box costs \$2.50 per square foot, and the four sides cost \$3.25 per square foot. The eight seams must be welded together at a cost of \$1.25 per foot. Find the dimensions and the cost of the most economical box. Show your work and explain how you got your answer.

2. **Distances between graphs**

   (a) Find the point on the graph of the function $f(x,y) = \dfrac{3}{xy}$ that is nearest to the point $(2,1,0)$. What is the distance? Show your work and explain how you got your answer. (Suggestion: Since distance is a positive function, you can solve this problem if you minimize the square of the distance.)

   (b) Find the minimum distance between the graph of $y = x^3$ and the graph of $y = \ln x$. Find the points on these curves where the minimum distance occurs and label them on the graphs of $y = x^2$ and $y = \ln x$. Show your work and explain how you got your answer.

3. **A television cable between two towns**: Town A is located 2 miles west of the west shore of the Mississippi river. Town B is located 7 miles east of the east shore and 2 miles north of Town A. The river itself is one half mile wide. A television cable is to be run from town A to town B. It costs $400 per mile to lay the cable on land and $900 per mile to lay the cable under water. Describe the path of the most economical cable and find its cost. Show your work and explain how you got your answer.

4. **Making wire figures**: A piece of wire 1 meter long is cut into three (or fewer) pieces. One is bent into the shape of a circle, another into an equilateral triangle, and another into a square. Explain how this should be done so that the sum of the areas enclosed by the three figures is (a) a minimum and (b) a maximum. (<u>Note</u>: In maximizing or minimizing the total area, you should allow for the possibilities that you may cut the wire into only two pieces and bend them into two of the above shapes, or perhaps not cut it at all and just bend the whole wire into one figure.)

Chapter 15

Double Integrals and Line Integrals

New *DERIVE* topics	•Calculating iterated integrals
Calculus concepts	•Double integrals •Double Riemann sums •Line integrals •Green's theorem

Let R denote the rectangle consisting of all points (x, y) satisfying $a \leq x \leq b$ and $c \leq y \leq d$. The double integral of f over R is defined as a limit of Riemann sums. *In this book we will always assume that the sum is taken for partitions of n equal subintervals on each axis.* Thus we have $\int\int_{\text{R}} f(x, y)\, dA = \lim_{n \to \infty} \sum_{i=1}^{n} \sum_{j=1}^{n} f(x_i^*, y_j^*) \Delta A$, where $\Delta A = \frac{(b-a)(d-c)}{n^2}$, and (x_i^*, y_j^*) is in the corresponding subrectangle.

Solved Problem 15.1: Double Riemann Sums

New *DERIVE* Lessons •Calculating iterated integrals

Let R be the rectangle defined by $1 \leq x \leq 3$, $2 \leq y \leq 5$.

(a) Approximate $\int\int_{\text{R}} xy\, dA$, using a double Riemann sum with 10 subintervals on each axis, taking (x_i^*, y_j^*) to be the upper right corner of each subrectangle.

(b) Calculate the double Riemann sum in Part (a) using n subintervals, taking (x_i^*, y_j^*) to be the upper right corner of each subrectangle. Calculate the double integral by taking the limit as n goes to ∞ of the result.

(c) Write the double integral as an iterated integral and use *DERIVE* to calculate it.

Solution to (a): We partition the intervals $1 \leq x \leq 3$ and $2 \leq y \leq 5$ into 10 subintervals so that $\Delta x = \frac{2}{10}$, $\Delta y = \frac{3}{10}$, and $\Delta A = \Delta x \Delta y = \frac{3}{50}$. Since we are using the upper right corner of each subrectangle, we have $x_i^* = 1 + i\frac{2}{10}$, and $y_j^* = 2 + j\frac{3}{10}$. Thus the sum we want is $\sum_{i=1}^{10} \sum_{j=1}^{10} f\left(1 + i\frac{2}{10}, 2 + j\frac{3}{10}\right)\frac{3}{50}$. First **Author** `F(x,y):=xy` as seen in expression 1 of Figure 15.1. Now **Author** `F(1+i(2/10), 2+j(3/10))(3/50)` and then use **Calculus Sum** with `variable:j` and `Lower limit:1` `Upper limit:10`. This sum appears in expression 3 of Figure 15.1. To make the double sum, use **Calculus Sum** again with `variable:i` and `Lower limit:1` `Upper limit:10`.

```
1:    F(x, y) := x y

              2          3    3
2:    F[1 + i ——, 2 + j ——] ——
              10         10   50

      10          2          3    3
3:    Σ  F[1 + i ——, 2 + j ——] ——
      j=1         10         10   50

      10  10          2          3    3
4:    Σ   Σ  F[1 + i ——, 2 + j ——] ——
      i=1 j=1         10         10   50

5:    45.99

              2         3    6
      F[1 + i —, 2 + j —] ——
6:            n         n    2
                             n

      n           2          3    6
      Σ  F[1 + i —, 2 + j —] ——
7:    j=1         n         n    2
                                  n
```

Figure 15.1: **A double sum using 10 subintervals**

We **approX** expression 4 to get the answer, 45.99, in expression 5.

Solution to (b): This time we want the sum using n subintervals. Thus $\Delta x = \frac{2}{n}$, $\Delta y = \frac{3}{n}$, $\Delta A = \frac{6}{n^2}$, $x_i^* = 1 + i\frac{2}{n}$, and $y_j^* = 2 + i\frac{3}{n}$. The double sum is $\sum_{i=1}^{n}\sum_{j=1}^{n} f\left(1 + i\frac{2}{n}, 2 + j\frac{3}{n}\right)\frac{6}{n^2}$. **Author** `F(1+i(2/n), 2+j(3/n))(6/n^2)` as seen in expression 6 of Figure 15.1. Now, make the double sum as in Part (a), but for each sum use Lower limit:1 Upper limit:n. The completed sum is expression 8 of Figure 15.2. We **Simplify** to get expression 9. Finally, we use **Calculus Limit** to calculate the limit as seen in expressions 10 and 11. The answer, 42 in expression 11, is the exact value of the double integral.

Solution to (c): You should be able to do this easily with pencil and paper, but we want to demonstrate how to do iterated integrals with *DERIVE*. The iterated integral may be expressed as either $\int_1^3 \int_2^5 xy\ dy\ dx$ or $\int_2^5 \int_1^3 xy\ dx\ dy$. To do the first, **Author xy** and use **Calculus Integrate** with variable:y and Lower limit:2 Upper limit:5. The integral is in expression 13 of Figure 15.2. With this expression highlighted, use **Calculus Integrate** again with variable:x and Lower limit:1 Upper limit:3. The iterated integral is in expression 14. When we **Simplify**, we get the same answer that we got in Part (b).

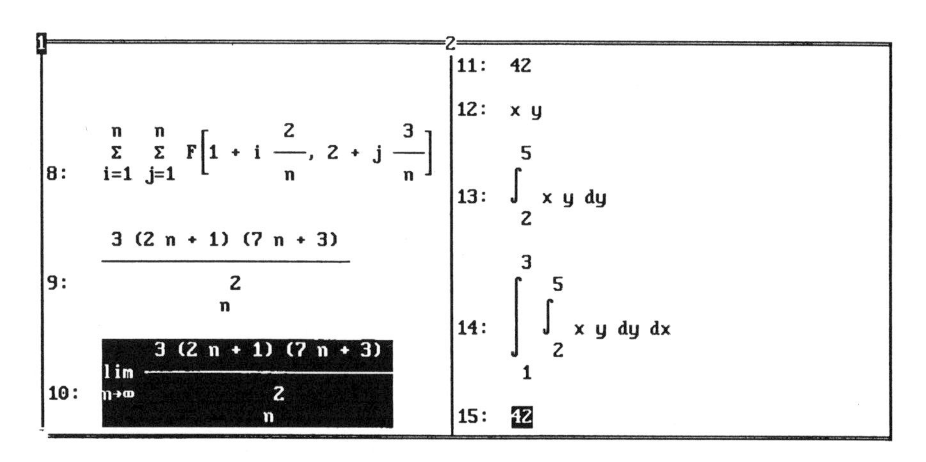

Figure 15.2: **The double integral as a limit of a double Riemann sum**

Solved Problem 15.2: Calculating iterated integrals

New *DERIVE* Lessons •Approximating iterated integrals

Calculate the iterated integrals (a) $\int_0^1 \int_y^1 ye^{x^3}\,dx\,dy$ and (b) $\int_0^1 \int_0^{\sqrt{1-y^2}} \frac{1}{1+x^2+y^2}\,dx\,dy$.

Solution to (a): **Author** y Alt e ^(x^3) as seen in expression 1 of Figure 15.3, and use **Calculus Integrate** with variable:x and Lower limit:y Upper limit:1. With expression 2 highlighted, use **Calculus Integrate** with variable:x and Lower limit:0 Upper limit:1. If you **Simplify**, *DERIVE* will return the integral unevaluated (try it), but just as with one-variable integrals, *DERIVE* can approximate iterated integrals that it cannot evaluate exactly. If we **approX** expression 3 of Figure 15.3, *DERIVE* returns the value 0.286380 in expression 4.

In this case, there is a way to check *DERIVE's* answer. We note first that the iterated integral is the double integral over the triangular region bounded by the x-axis, the line $x = 1$, and the graph of $y = x$. (You are encouraged to draw a picture.) Thus if we change the order of integration, we get $\int_0^1 \int_y^1 ye^{x^3}\,dx\,dy = \int_0^1 \int_0^x ye^{x^3}\,dy\,dx$. This new integral appears in expression 6 of Figure 15.3, and with the integral expressed in this way, *DERIVE* can **Simplify** to obtain the answer in expression 7. When we **approX**, we see from expression 8 that *DERIVE's* original approximation was very accurate.

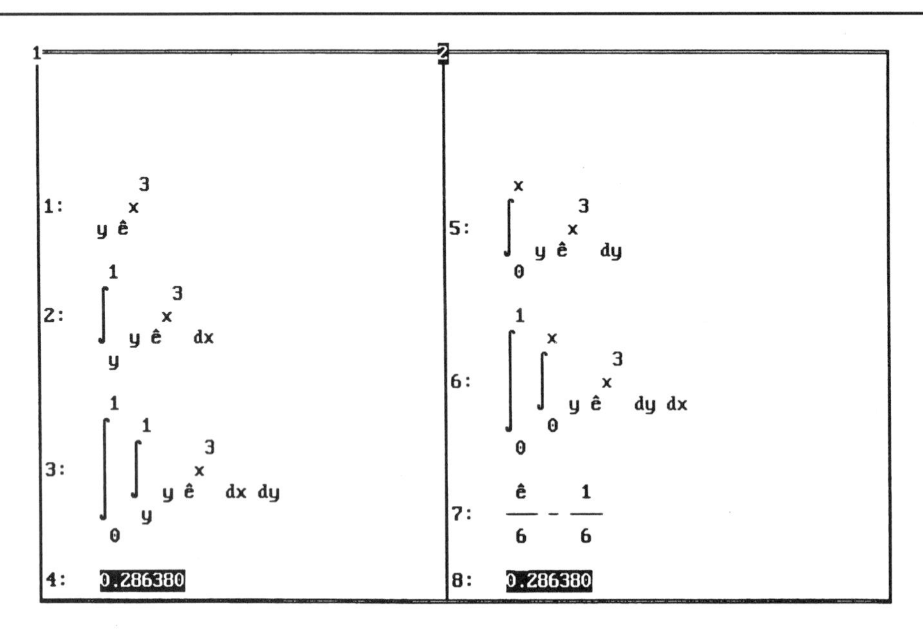

Figure 15.3: **Changing the order of integration**

Solution to (b): The iterated integral appears in expression 11 of Figure 15.4. When we **Simplify**, we see from expression 12 that *DERIVE* is able to evaluate the inside integral, but cannot complete the calculation. We **approX** to get the approximation in expression 13. Once again, there is a way to check *DERIVE's* answer. The iterated integral is the double integral over the quarter of the unit disk that lies in the first quadrant. (You are encouraged to draw a picture.) If we change to polar coordinates, the integrand becomes $\frac{1}{1+r^2}$. The quarter disk is described by $0 \leq r \leq 1$ and $0 \leq \theta \leq \frac{\pi}{2}$. Thus we get $\int_0^{\frac{\pi}{2}} \int_0^1 \frac{1}{1+r^2} r \, dr \, d\theta$. This integral and its value appear in expressions 16 and 17 of Figure 15.4. When we **approX** expression 17, we see that *DERIVE's* original approximation of the integral was accurate to four places.

Notation for line integrals: If $\mathbf{F}(x,y) = [g(x,y), h(x,y)]$, and if $\mathbf{r}(t)$, with $a \leq t \leq b$, is a parameterization of the curve C, then the line integral of $\mathbf{F}$ over C is commonly written in three equivalent ways: $\int_C \mathbf{F} \cdot d\mathbf{r}$, $\int_C g(x,y) \, dx + h(x,y) \, dy$, or $\int_a^b \mathbf{F}(\mathbf{r}(t)) \cdot \mathbf{r}'(t) \, dt$. We will use the first notation, but you should be familiar with all three.

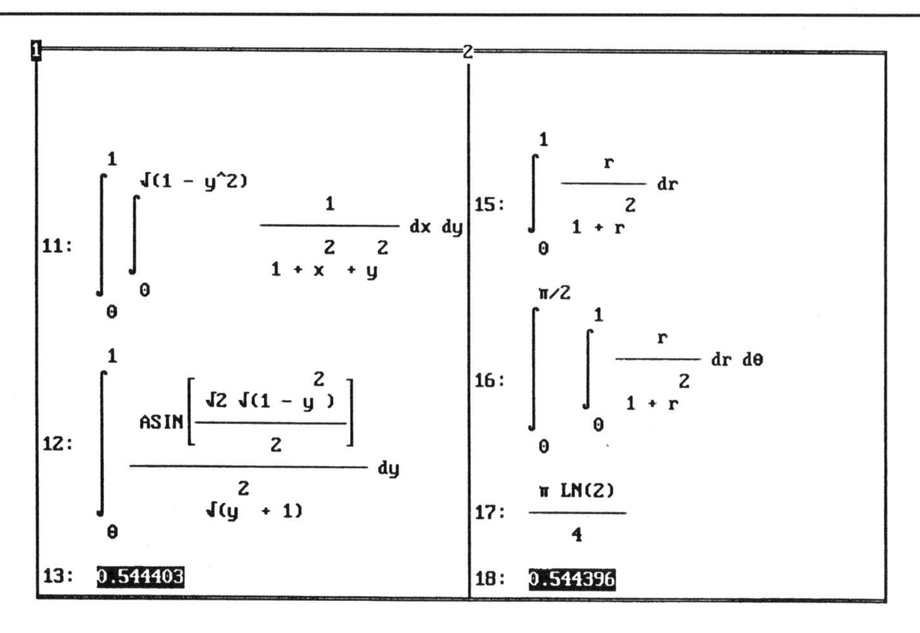

Figure 15.4: **Changing to polar coordinates**

Solved Problem 15.3: Line integrals

Let C be the first quarter of the unit circle oriented counterclockwise, and let $\mathbf{F}(x, y) = [xy^2, x^3 - y^3]$. Calculate the line integral of **F** over C.

Solution: We will first show the steps in evaluating the line integral and then present a function that automates the procedure. First **Author** `[xy^2,x^3-y^3]` as seen in expression 1 of Figure 15.5. Next we use **Manage Substitute** to replace x by `COSt` and y by `SINt`, the standard parameterization of the unit circle. The result is in expression 2 of Figure 15.5. Now **Author** `[COSt, SINt]` and use **Calculus Differentiate** to get the derivative in expression 5. For the integrand, we **Author** `#2.#5` and **Simplify**. Finally use **Calculus Integrate** with Lower limit:`0` Upper limit:`pi/2`. **Simplify** to get the value of the line integral in expression 9 of Figure 15.5.

You should understand the steps in constructing a line integral, but we will make a function that automates all this. Begin with a clear screen and **Author** the following expressions exactly as they appear.

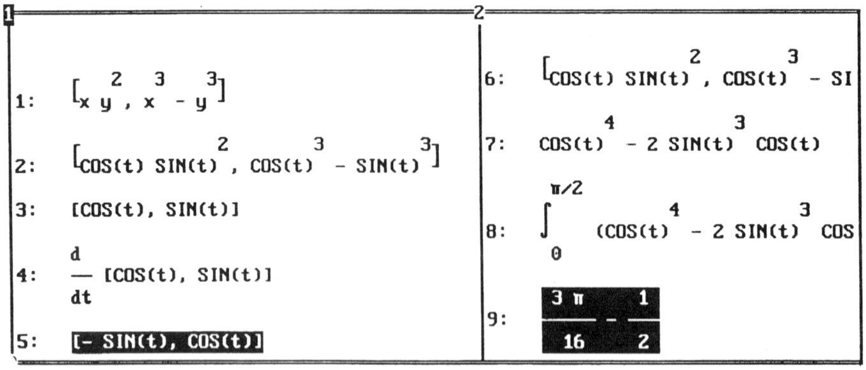

```
1:  [x y^2, x^3 - y^3]

2:  [COS(t) SIN(t)^2, COS(t)^3 - SIN(t)^3]

3:  [COS(t), SIN(t)]

     d
4:  -- [COS(t), SIN(t)]
    dt

5:  [- SIN(t), COS(t)]

6:  [COS(t) SIN(t)^2, COS(t)^3 - SI

7:  COS(t)^4 - 2 SIN(t)^3 COS(t)

     π/2
8:  ∫     (COS(t)^4 - 2 SIN(t)^3 COS
     0

     3 π   1
9:  ---- - -
     16    2
```

Figure 15.5: **Evaluating a line integral**

```
LIM(f, [x,y], r)
```

```
LINEINT(f,r,a,b):=INT(#1.DIF(r,t),t,a,b)
```

You may wish to save this as a file using **Transfer Save Derive** LINEINT. You can retrieve it using **Transfer Merge** LINEINT. In this file f plays the role of $\mathbf{F}(x, y)$ and r plays the role of $\mathbf{r}(t)$. To use the LINEINT function for this problem, we **Author** `LINEINT([xy^2, x^3-y^3],[COSt, SINt],0,pi/2)` and **Simplify** to get the answer in expression 4 of Figure 15.6.

Solved Problem 15.4: Green's Theorem

Verify Green's theorem for $\int_{C}[x^3y^2, xy^2] \cdot d\mathbf{r}$, where C is the boundary of the ellipse $\frac{x^2}{3} + \frac{y^2}{5} = 1$ oriented counterclockwise.

Solution: Load the LINEINT.MTH file using **Transfer Merge** LINEINT. We can parameterize the ellipse using $\mathbf{r}(t) = [\sqrt{3}\cos t, \sqrt{5}\sin t]$, $0 \le t \le 2\pi$. (You should verify that this is a correct parameterization by showing that $x = \sqrt{3}\cos t$, $y = \sqrt{5}\sin t$ satisfies the equation for the ellipse and by plotting the graph.) Since we will later need the individual coordinates of $\mathbf{F}$, we **Author** `x^3y^2` and then `xy^2` separately as expressions 3 and 4 of Figure 15.7.

Author `LINEINT([#3, #4],[SQRT(3)COSt, SQRT(5)SINt],0,2 pi]` and **Simplify** to get the line integral as seen in expression 6 of Figure 15.7.

```
1:   lim    f
     [x, y]→r
```

$$\text{2:}\quad \text{LINEINT}(f, r, a, b) := \int_a^b \left(\lim_{[x,y]\to r} f\right) \cdot \frac{d}{dt} r \, dt$$

$$\text{3:}\quad \text{LINEINT}\left[\left[x\, y^2,\ x^3 - y^3\right],\ [\text{COS}(t),\ \text{SIN}(t)],\ 0,\ \frac{\pi}{2}\right]$$

$$\text{4:}\quad \frac{3\pi}{16} - \frac{1}{2}$$

Figure 15.6: **Automating the calculation of line integrals**

To verify Green's theorem, we must calculate the double integral over the ellipse of $\frac{\partial}{\partial x} xy^2 - \frac{\partial}{\partial y} x^3y^2$. We **Author** `DIF(#4,x)-DIF(#3,y)` and **Simplify** to get the integrand (not shown in Figure 15.7). The region enclosed by the ellipse is described by $-\sqrt{3} \leq x \leq \sqrt{3}$ and $-\sqrt{5(1 - \frac{x^2}{3})} \leq y \leq \sqrt{5(1 - \frac{x^2}{3})}$. Thus the integral we want is

$$\int_{-\sqrt{3}}^{\sqrt{3}} \int_{-\sqrt{5(1 - \frac{x^2}{3})}}^{\sqrt{5(1 - \frac{x^2}{3})}} y^2 - 2x^3y \, dy \, dx$$

It appears in expression 10 of Figure 15.7, and when we **Simplify**, we get the same answer as before in expression 11.

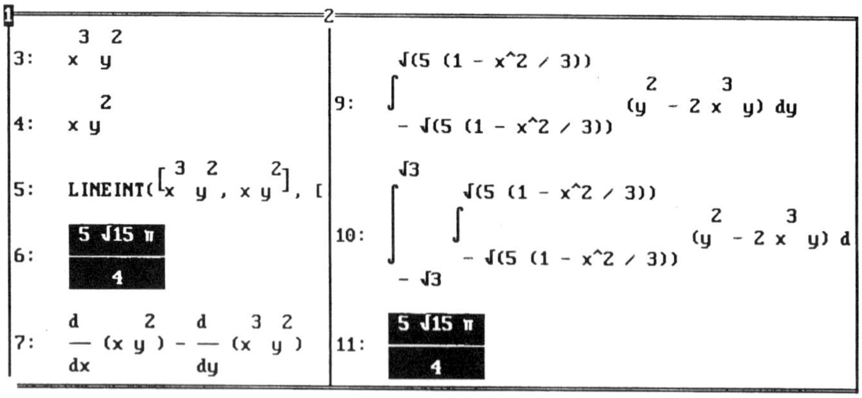

```
3:   x^3 y^2

4:   x y^2

5:   LINEINT([x^3 y^2, x y^2], [

6:   5 √15 π / 4

7:   d/dx (x y^2) - d/dy (x^3 y^2)

9:   ∫_{- √(5 (1 - x^2 / 3))}^{√(5 (1 - x^2 / 3))} (y^2 - 2 x^3 y) dy

10:  ∫_{- √3}^{√3} ∫_{- √(5 (1 - x^2 / 3))}^{√(5 (1 - x^2 / 3))} (y^2 - 2 x^3 y) d

11:  5 √15 π / 4
```

Figure 15.7: $\int_C \mathbf{F} \cdot d\mathbf{r}$

Practice Problems

1. Use *DERIVE* to calculate $\int_0^{\pi} \int_0^{\pi} \sin(x+y)\, dy\, dx$. Answer: 0
2. Approximate $\int_0^1 \int_0^1 e^{x^3+y^3}\, dy\, dx$. Answer: 1.80070
3. Calculate the line integral of $\mathbf{F}(x,y) = [x^2y^3, x^3y^2]$ over the curve $\mathbf{r}(t) = [\cos(2t), \sin(3t)]$, $0 \le t \le \frac{\pi}{2}$. We suggest you calculate the integral showing each of the steps and then use the LINEINT.MTH file to check your answer. Answer: $\frac{1}{3}$

Laboratory Exercise 15.1

Double Integrals, Iterated Integrals, and Double Riemann Sums

Name ______________________________ **Due Date** __________

Let R denote the rectangle $1 \leq x \leq 4$, $-1 \leq y \leq 1$ and let (x_i^*, y_j^*) be the upper right corner of each subrectangle in the double Riemann sums in this exercise.

1. Approximate $\int\int_R x^2 y^2 \, dA$ using a double Riemann sum with 10 subintervals on each axis.

2. Repeat Part 1 using 100 subintervals.

3. Repeat Part 1 using n subintervals and calculate the limit as n goes to ∞ of the result.

4. Comment on the relative accuracy of the approximations in Parts 1, 2, and 3.

5. Write the double integral as an iterated integral and use *DERIVE* to calculate it. Compare your answer with the one you got in Part 3.

6. Plot the graph of x^2y^2 and explain what this double integral means in terms of volume.

Laboratory Exercise 15.2

Evaluating Iterated Integrals

Name ______________________________ **Due Date** ____________

For each of the following integrals, (a) use *DERIVE* to approximate its value, (b) draw a picture of the region you are integrating over, and (c) follow the given suggestion to evaluate the integral exactly.

1. $\int_0^1 \int_{y^{\frac{1}{3}}}^1 y^2 \sin(x^2)\, dx\, dy$ (Suggestion for (c): Reverse the order of integration.)

2. $\int_{-2}^2 \int_0^{\sqrt{4-y^2}} \sin(x^2 + y^2)\, dy\, dx$ (Suggestion for (c): Change to polar coordinates.)

3. $\displaystyle\int_{-3}^{3}\int_{-\sqrt{9-y^2}}^{\sqrt{9-y^2}} \frac{x}{2+x^2+y^2}\,dx\,dy$ (Suggestion for (c): Change to polar coordinates.)

4. $\displaystyle\int_{0}^{1}\int_{y^3}^{1} \sqrt{xy}\cos x\,dx\,dy$ (Suggestion for (c): Change the order of integration.)

Laboratory Exercise 15.3

Applications of Double Integrals

Name ______________________ **Due Date** ____________

In each of the following problems express the answer as a double integral, write it as an iterated integral, and evaluate or approximate it. Some of them require formulas that are not given here, but should be in your text.

1. Let R be the region in the first quadrant that is enclosed by the graphs of $y = \sin x$ and $y = \dfrac{x}{2}$. Find the volume of the solid that lies beneath the graph of $z = \sqrt{1 + x^2 + y^2}$ and above the region R.

2. The unit sphere, $x^2 + y^2 + z^2 = 1$, is cut in two pieces by the plane $z = \frac{1}{2}$. Find the centroid of each piece.

3. Find the centroid of the solid that lies beneath the surface $z = e^{3x+2y-x^2-y^2}$ and above the xy-plane.

4. Let R be the unit disk in the xy-plane. Find the area of the portion of the surface $z = e^{-x^2y^2}$ that lies above R.

Laboratory Exercise 15.4

A Conservative Vector Field

Name ________________________________ **Due Date** ____________

Consider $\mathbf{F}(x, y) = [ye^{xy}, xe^{xy}]$.

1. Let $\mathbf{r}(t) = [t, t]$, $0 \le t \le 1$. Plot the graph of $\mathbf{r}$ and calculate the line integral of $\mathbf{F}$ over $\mathbf{r}$.

2. Let $\mathbf{s}(t) = [t^3, t^2]$, $0 \le t \le 1$. Plot the graph of $\mathbf{s}$, compare it with $\mathbf{r}$, and calculate the line integral of $\mathbf{F}$ over $\mathbf{s}$.

3. Let $\mathbf{u}(t) = [\sin(6\pi t) + \sin(\frac{\pi}{2}t), t\cos(2\pi t)]$, $0 \le t \le 1$. Plot the graph of $\mathbf{u}$, compare it with the graphs of $\mathbf{r}$ and $\mathbf{s}$, and calculate the line integral of $\mathbf{F}$ over $\mathbf{u}$.

4. Verify that $\mathbf{F}(x, y)$ is a *conservative vector field*; that is, $\dfrac{\partial}{\partial y} ye^{xy} = \dfrac{\partial}{\partial x} xe^{xy}$.

5. Find a function $g(x, y)$ such that $\nabla g = \mathbf{F}$.

6. Use your answer from Part 5 to calculate the line integral of $\mathbf{F}$ over any curve C that begins at the origin and ends at $(1, 1)$. Does this explain your answers in Parts 1, 2, and 3?

7. Let C be any curve that begins at $(3, 6)$ and ends at $(3, 6)$. Without doing any calculation, explain how you can evaluate the line integral of $\mathbf{F}$ over the C.

Laboratory Exercise 15.5

Calculating Work

Name ______________________________ Due Date ____________

Let $\mathbf{F}(x, y) = [y^4 - y^2 - 3x^2, x^2 - 3y^3]$.

1. Verify that for any constant, d, the curve $\mathbf{r}(t) = [t, d(t - t^2)]$, $0 \leq t \leq 1$ begins at the origin and ends at $(1, 0)$. Calculate the work in moving from the origin to $(1, 0)$ along the path $\mathbf{r}$ under the influence of the force field $\mathbf{F}$.

2. Find all values of d so that the work from Part 1 is 0.

3. Find a value of d so that the work from Part 1 is a minimum.

4. Can you find a value of d so that the work from Part 1 is a maximum? Explain.

Laboratory Exercise 15.6

Using Green's Theorem

Name ______________________________ **Due Date** ____________

1. Evaluate the line integral of $[e^{x^3} + y^4, \sin y - x^3]$ counterclockwise around the unit circle.

2. Apply Green's theorem to calculate the line integral in Part 1.

3. (This part does not use the computer.) Use Green's theorem to show that if R is a region whose boundary is the closed curve C oriented in the counterclockwise direction, then the area of R is given by the line integral over C of $[-y, x]$.

4. Use the result in Part 3 to verify that the area of a circle of radius r is πr^2.

5. Plot the graph of $[\cos^3 t, \sin^5 t]$, $0 \leq t \leq 2\pi$, and use the result from Part 3 to calculate the area enclosed by the curve.

Chapter 16

Differential Equations

New *DERIVE* topics •Solving differential equations •Solving initial value problems •Finding approximate solutions •Plotting approximate solutions •Solving second order equations
Calculus concepts •Differential equation •Initial value problem •Euler's method •Second order linear equation

A Note about the ODE1.MTH File

In versions of *DERIVE* prior to 2.6, there is an error in the ODE1.MTH file that is used to solve differential equations. If you have an early version of *DERIVE* the `DSOLVE1_GEN` and `DSOLVE1` functions in this file report "inapplicable" to equations that *DERIVE* knows how to solve.

Solved Problem 16.1: Solving first order equations

New *DERIVE* Lessons •Solving differential equations •Solving initial value problems

Consider the differential equation $xy' + 2y = e^{-x}$.

(a) Use *DERIVE* to find the general solution of the equation and check the answer.

(b) Plot the graphs of several members of the family of solutions.

(c) Find the solution subject to the initial condition $y(1) = 3$.

Solution to (a): In order to use *DERIVE* to solve differential equations, we must first load the utility file ODE1.MTH that comes with the *DERIVE* program. We will use only two of the functions in the file, so to save time, we load the file into memory rather than onto the screen. Use **Transfer Load Utility** ODE1. Nothing will appear on the screen, but the file is now loaded into memory.

We want to use the function DSOLVE1_GEN, but its name and function arguments are a bit unhandy to type in, so we will begin by shortening its name. To name it D, we **Author** `D(p,q):=DSOLVE1_GEN(p,q,x,y,c)` (not shown in Figure 16.1). Now, for a differential equation of the form $p + qy' = 0$, $D(p, q)$ will attempt to find the general solution. To solve our equation,

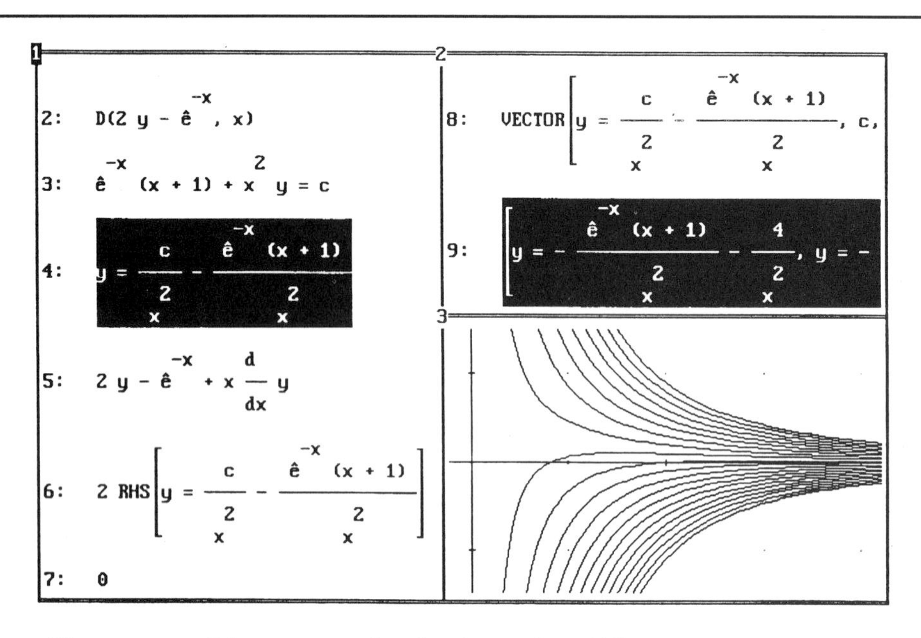

Figure 16.1: **The general solution of a differential equation**

we must first write it in this form, $(2y - e^{-x}) + xy' = 0$. Thus we **Author** `D(2y-`Alt e`^-x, x)` and **Simplify** to get the solution in expression 3. (We should note that neither *DERIVE* nor any other piece of software can solve all first order differential equations. If *DERIVE* had been unable to find the solution, it would have returned the message "inapplicable.") In this case, we can **soLve** expression 3 setting SOLVE variable:y to get the solution explicitly as a function of x in expression 4.

To check the solution, we first enter the left-hand side of the differential equation as written above. **Author** 2y-Alt e`^-x+xDIF(y,x)`. The result is in expression 5 of Figure 16.1. Use **Manage Substitute**, and at the prompt MANAGE SUBSTITUTE value:y type `RHS(#4)` to indicate that the right-hand side of expression 4 is to be plugged in for y. The result is partially displayed in expression 6, but when we **Simplify**, we get 0 and conclude that expression 4 really is a solution of the equation.

Solution to (b): We will make a picture of the solution curves as c ranges from -4 to 4 in steps of 0.4. Thus we **Author** `VECTOR(#4,c,-4,4,0.4)`, **approX**, **Plot**, and wait a while to see all the curves in window 3 of Figure 16.1. We emphasize that each of the graphs in the picture (as well as infinitely many others) is a solution of the equation.

Solution to (c): One method of solving this part of the problem is to plug $x = 1$ and $y = 3$

in expression 4 and then **soLve** for c, but we want to illustrate the use of the DSOLVE1 function in the ODE1.MTH file that is designed to give solutions of differential equations with given initial conditions. Once again we will shorten its name. **Author** `DS(p,q,a,b):=DSOLVE1(p,q,x,y,a,b)` as seen in expression 10 of Figure 16.2. Now $D(p,q,a,b)$ will attempt to solve the initial value problem $p+qy'=0$ with $y(a)=b$. (That is $y=b$ when $x=a$.) In our case, we want $y(1)=3$ so we **Author** `DS(2y-`[Alt e]`^-x, x, 1, 3)`. **Simplify** and then **soLve** for y to get the solution in expression 13.

```
10:  DS(p, q, a, b) := DSOLVE1(p, q, x, y, a, b)

                  -x
11:  DS(2 y - ê   , x, 1, 3)

          -1     -x              2
12:  - 2 ê    + ê   (x + 1) + x  y - 3 = 0

             -1        -x
          2 ê         ê    (x + 1)      3
13:  y = ------- - ---------------- + ----
             2             2            2
            x             x            x
```

Figure 16.2: **The solution of a differential equation with initial conditions**

Solved Problem 16.2: Baking potatoes

At time $t=0$, a potato is placed in an oven that has been preheated to a temperature of 400 degrees Fahrenheit. The potato had an initial temperature of 75 degrees Fahrenheit. The potato heats according to Newton's law so that the rate of change in the temperature of the potato is proportional to the difference in the temperature of the potato and that of the oven. At time $t=20$ minutes, the temperature of the potato was 180 degrees. The potato will be done when it reaches a temperature of 270 degrees. When should it be removed from the oven?

Solution: Let x denote the time variable and y the temperature of the potato. Then according to Newton's law, $\frac{dy}{dx}=\alpha(400-y)$ where α is the constant of proportionality. You should be able to solve this equation by hand, but we will show how *DERIVE* handles the problem. We assume that the ODE1.MTH file has been loaded into memory and the DS function has been defined according to the instructions in Solved Problem 16.1. We write our equation in the form $\alpha(400 - y) - y' = 0$. According to the initial condition, $y = 75$ when $x = 0$. Thus we **Author** `DS(alpha(400-y),-1,0,75)` which appears in expression 2 of Figure 16.3. (Note: As you can see, we have made the Greek letter α by typing `alpha`. You can also make it using Alt a, or you can use some other letter if you prefer.) When we **Simplify**, we see in expression 3 that *DERIVE* has introduced the complex number $i = \sqrt{-1}$ into the solution, but when we **soLve** this for y to get the temperature as a function of time, we see in expression 4 that it goes away. You should not be concerned with the appearance of the number i; it arises because *DERIVE* uses some advanced methods for handling logarithms.

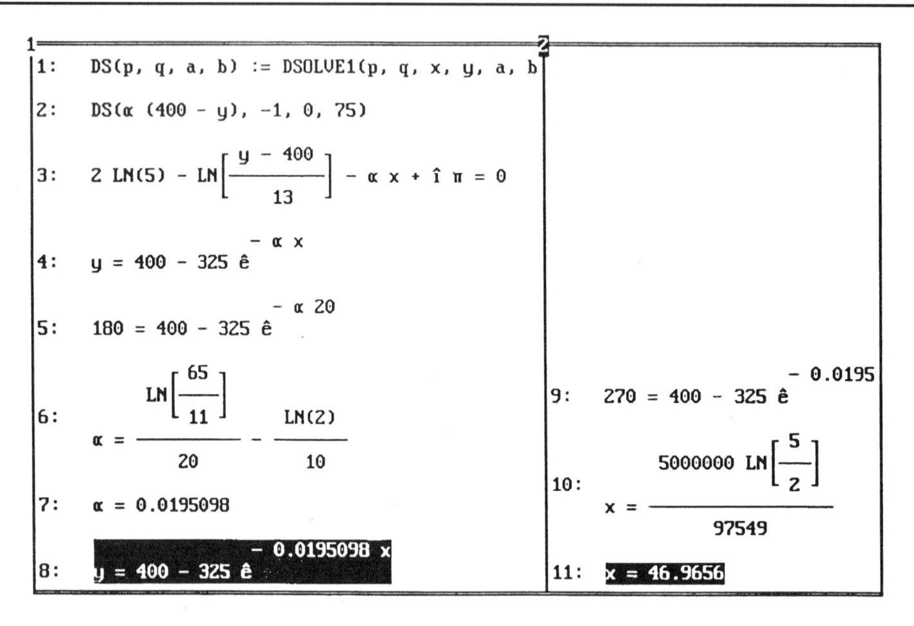

Figure 16.3: **Finding when a potato is done**

Before we can finish the problem, we need to use the fact that $y = 180$ when $x = 20$ to figure out what the value of α is. We highlight expression 4 of Figure 16.3 and use **Manage Substitute** to replace x by 20 and y by 180. Next we **soLve** expression 5 and then **approX** to get the value of α in expression 7. We put this into expression 4 to get the equation for the temperature as a function of time in expression 8. Finally, we use **Manage Substitute** to replace y in expression 8 by the temperature of the cooked potato, 270, and **soLve** to get the time the potato should be removed from the oven in expression 11.

Solved Problem 16.3: Approximating solutions of differential equations

New *DERIVE* Lessons •Using the EULER function •Plotting approximate solutions

Find an approximate solution of the initial value problem $yy' + \sqrt{y} = x$, $y(1) = 2$. What is the approximate value of $y(3)$? Plot the graph of the approximate solution.

Solution: If you try to get the exact solution of the equation using `D(SQRT(y)-x,y)`, *DERIVE* will return the message "inapplicable" indicating that it does not know how to find the exact solution. Thus we must settle for an approximate solution. Load the file ODE_APPR.MTH into memory using **Transfer Load Utility** ODE_APPR.MTH. We will make use of the function EULER$(r, x, y, x0, y0, h, n)$ whose arguments are described below.

- The differential equation must be in the form $y' = r$.
- x is the independent variable, and y is the dependent variable.
- $(x0, y0)$ is your starting point. That is $y(x0) = y0$.
- h is the step size.
- n is the number of steps you wish to execute.

Your calculus text may or may not include a discussion of Euler's method, and so a few words about h and n are in order. Basically, Euler's method begins at the starting point $(x0, y0)$ and approximates the value of $(x0+h, y(x0+h))$, then $(x0+2h, y(x0+2h))$, and so on until it gets to $(x0+nh, y(x0+nh))$. Smaller values of the step size h generally yield more accurate answers at the expense of speed of calculation and available memory. The solution of this problem should clarify things a bit.

We first write the equation in the form $y' = \dfrac{x - \sqrt{y}}{y}$. Our starting point is $x0 = 1$, $y0 = 2$. We will use a step size of $h = 0.1$, and we want a value for $y(3)$. Thus we must choose n to be at least $\dfrac{3-1}{0.1} = 20$. We **Author** `EULER((x-SQRT(y))/y, x, y, 1, 2, 0.1, 20)` as seen in expression 1 of Figure 16.4. When we **approX** (do not **Simplify**!), *DERIVE* gives the list of approximate function values partially displayed in expression 2. If we use the right arrow key , [→], to move out to the end of expression 2, we see that $y(3)$ is approximately 2.46039. To check for accuracy, we will repeat this using a step size of $h = 0.05$. Since we want to get a look at $y(3)$, this will require $n = 40$. Thus we **Author** `EULER((x-SQRT(y))/y,x,y,1,2,0.05,40)` as seen in expression 3 of Figure 16.4. When we **approX**, we get expression 4. This time we

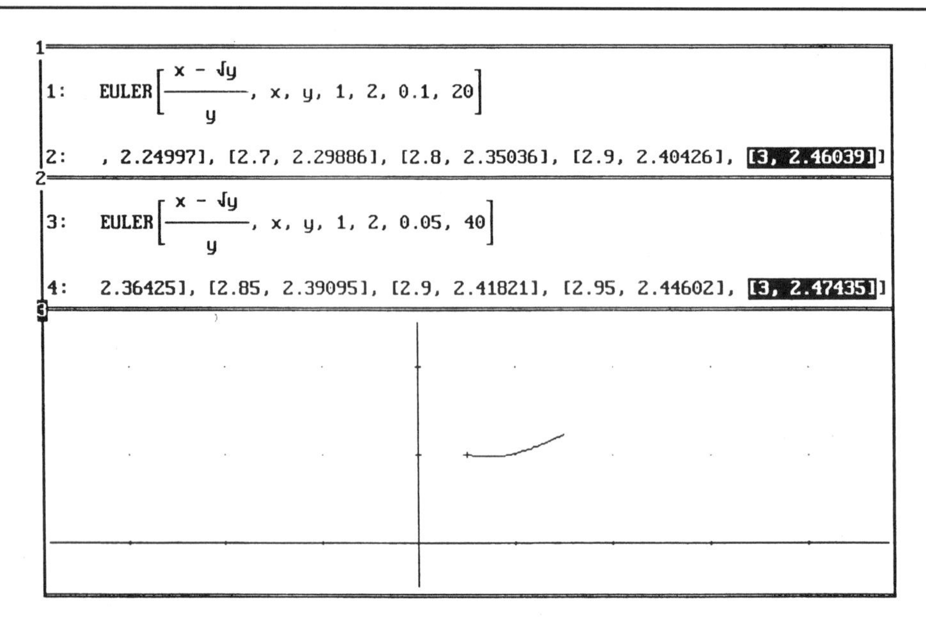

Figure 16.4: **Using Euler's method to approximate solutions, Scale x:2 y:2**

get an approximate value of $y(3) = 2.47435$. Because the step size is smaller, we expect this answer to be more accurate than the first one.

To plot the graph we highlight expression 4 of Figure 16.4 and **Plot** to open a plot window. To make the graph smooth, we use **Options State Rectangular Connected Small**. Now **Plot** again to get the graph in window 3.

Solved Problem 16.4: Solving second order linear equations

New *DERIVE* Lessons •Solving second order equations

(a) Use *DERIVE* to find the general solution of $3y'' + 9y' + 6y = \sin x$ and check the answer.

(b) Solve the equation in (a) if the initial values, $y(0) = 1$, $y'(0) = 0$ are added.

Solution to (a): To solve second order equations with *DERIVE* we must first load the ODE2.MTH file into memory using **Transfer Load Utility** ODE2. To find the solution of $3y'' + 9y' + 6y = \sin x$, we use DSOLVE2$(p, q, r, x, c1, c2)$. This function is designed to solve the differential equation $y'' + p(x)y' + q(x)y = r(x)$ with $c1$ and $c2$ naming the arbitrary constants in the solution. Notice that the coefficient on y'' is assumed to be 1 here, and so we will first write our equation in the form

$$y'' + 3y' + 2y = \frac{\sin x}{3}$$

Now, to solve the equation, we **Author** and **Simplify** `DSOLVE2(3,2,SINx/3,x,c1,c2)`. The result is in expression 2 of Figure 16.5.

2: $\quad c1\ \hat{e}^{-x} + c2\ \hat{e}^{-2\ x} - \frac{\text{COS}(x)}{10} + \frac{\text{SIN}(x)}{30}$

3: $\quad \left[\frac{d}{dx}\right]^2 y + 3\ \frac{d}{dx}\ y + 2\ y$

4: $\quad \left[\frac{d}{dx}\right]^2 \left[c1\ \hat{e}^{-x} + c2\ \hat{e}^{-2\ x} - \frac{\text{COS}(x)}{10} + \frac{\text{SIN}(x)}{30}\right] + 3\ \frac{d}{dx} \left[c1\ \hat{e}^{-x} + c2\ \hat{e}^{-2\ x}\right.$

5: $\quad \frac{\text{SIN}(x)}{3}$

6: $\quad \text{DSOLVE2\_IV}\left[3,\ 2,\ \frac{\text{SIN}(x)}{3},\ x,\ 0,\ 1,\ 2\right]$

7: $\quad \frac{25\ \hat{e}^{-x}}{6} - \frac{46\ \hat{e}^{-2\ x}}{15} - \frac{\text{COS}(x)}{10} + \frac{\text{SIN}(x)}{30}$

Figure 16.5: **Solving a second order equation**

To check the answer, we enter the left-hand side of the equation as `DIF(y,x,2)+3DIF(y,x)+2y` as seen in expression 3. Now use **Manage Substitute** to replace y by **#2** and **Simplify**. We see from expression 5 that this is indeed the solution of the equation.

Solution to (b): To solve the initial value problem we use DSOLVE2_IV$(p, q, r, x, x0, y0, v0)$. This function attempts to solve the initial value problem $y'' + p(x)y' + q(x)y = r(x)$ with $y(x0) = y0$ and $y'(x0) = v0$. Thus we **Author** and **Simplify** `DSOLVE2_IV(3,2,SINx/3,x,0,1,2)`. The solution is in expression 7 of Figure 16.5.

Practice Problems

1. Use *DERIVE* to get the general solution of $y + \sin x + y' = 0$.
   Answer: $y = (x + c)\cos\left(\frac{\pi + 2x}{4}\right) - 1$

2. Use *DERIVE* to get the solution of $y \cot x + y' = \cot x$, with $y(3) = 5$.
   Answer: $y = \frac{0.564481}{\sin x} + 1$

3. Use the EULER function to approximate a solution of $y' = x^2 - y^2$ with $y(1) = 3$. Use $h = 0.1$ and $n = 50$. What is the approximate value of $y(6)$? Answer: $y(6) \approx 5.91484$

4. Use *DERIVE* to get the general solution of $y'' + y = \tan x$. Answer: $-\cos x \ln\left(\tan\left(\frac{\pi + 2x}{4}\right)\right) + c_1 \cos x + c_2 \sin x$.

5. Solve the equation in (4) subject to the initial values $y(0) = 0$ and $y'(0) = 1$. Answer: $2 \sin x - \cos x \ln\left(\tan\left(\frac{\pi + 2x}{4}\right)\right)$

Laboratory Exercise 16.1

Solving a First Order Equation

Name ______________________________ Due Date __________

Consider the differential equation $y + y' \cos x = \sin x$.

1. Find the general solution of the equation.

2. Check your answer in Part 1.

3. Solve this differential equation subject to the initial condition $y(1) = 2$.

4. Plot the graphs of at least 10 solution curves. Be sure to include your solution from Part 3 as one of the curves and label it.

Laboratory Exercise 16.2

Applications of First Order Equations

Name ______________________________ Due Date __________

1. **Baking potatoes quickly**: A chef realizes at 6:30 that she has forgotten to preheat the oven for baked potatoes to be served with dinner at 7:30. She turns on the oven and puts in the potatoes. The temperature of the oven t minutes after it is turned on is given by $O(t) = 400 - 325e^{-0.3t}$ degrees Fahrenheit. The potatoes heat according to Newton's law: The rate of change in the temperature, $P(t)$, of the potatoes is

$$\frac{dP}{dt} = 0.02(O(t) - P(t))$$

   (a) The initial temperature of the potatoes was 75 degrees Fahrenheit. (Explain how we know this from what is given.) Find the temperature of the potatoes as a function of time.

   (b) The potatoes will be done when they reach a temperature of 270 degrees. Will they be done in time for dinner?

   (c) If the oven had been properly preheated, then they would have heated according to $\frac{dP}{dt} = 0.02(400 - P(t))$. How much time did the chef lose by forgetting to preheat the oven?

2. **An equilibrium solution**: A balloon, which initially contains V_0 cubic inches of air, leaks air through a small hole at a rate equal to 4 percent of its volume at any given time. Additional air is being blown into the balloon at a constant rate of 25 cubic inches per minute.

   (a) Set up a differential equation whose solution gives the volume of air in the balloon as a function of time.

   (b) Solve the equation from Part 1. Express your answer in terms of t and V_0.

   (c) Find the limit as t goes to infinity of your solution. This is known as an *equilibrium solution* of the equation.

   (d) What value of V_0 gives the equilibrium solution?

   (e) Plot the graph of the equilibrium solution and at least five other solutions.

   (f) Explain in practical terms what the equilibrium solution means.

Laboratory Exercise 16.3

Euler's Method and the Step Size

Name ______________________________ Due Date ____________

Consider the initial value problem $xy + x^2y' = 1$, $y(1) = 2$.

1. Find the exact solution of the equation.

2. Use the EULER function with $h = 0.1$ and $n = 30$ to approximate the value of $y(4)$ and compare it with the exact value you get from Part 1.

3. Use the EULER function with $h = 0.05$ and $n = 60$ to approximate the value of $y(4)$ and compare it with the exact value you get from Part 1 and the approximate value you got from Part 2.

4. On the same screen, plot the graphs of the exact solution and the approximate solutions you found in Parts 2 and 3.

5. Explain how the step size, h, affects the accuracy of the EULER function for this differential equation.

Laboratory Exercise 16.4

Applications Using Euler's Method

Name ______________________________ Due Date __________

1. **Terminal velocity**: The velocity in feet per second of a falling object subject to air resistance is governed by the differential equation $\frac{dv}{dt} = 32 - 0.4v^{\frac{3}{2}}$. We take the initial velocity of the object to be 0.

   (a) Use *DERIVE* to try and get the exact solution of the equation. Report what happens.

   (b) Approximate the solution of the equation using the EULER function with $h = 0.01$ and $n = 300$, and plot the graph.

   (c) What is the approximate velocity at $t = 2$ seconds.

   (d) Use the graph and the data you have produced to estimate the *terminal velocity* of the object. That is what is the maximum velocity that the object can reach?

   (e) How long does it take the object to reach 90% of its terminal velocity?

2. **Population growth**: For a certain group of animals, the logistic model of population growth *with a threshold* states that the population, $P(t)$, as a function of time is governed by

$$\frac{dP}{dt} = -P\left(1 - \frac{P}{500}\right)\left(1 - \frac{P}{2000}\right)$$

(a) Use *DERIVE* to get the exact solution of the differential equation. What happens when you try to express P explicitly as a function of t?

(b) Use the EULER function with $h = 0.05$ and $n = 100$ to get approximate solutions of the equation in case the initial population is 400, 500, 600, 1000, 2000, 2500.

(c) On the same screen, plot the graphs of each of the solutions you found in (c).

(d) Explain what happens to populations if the initial population is less than 500, between 500 and 2000, and larger than 2000.

Laboratory Exercise 16.5

Solving Second Order Equations

Name ________________________ **Due Date** ____________

1. Consider the differential equation $y'' + y' + qy = 0$.

   (a) Solve the equation in case $q = -1$.

   (b) Solve the equation in case $q = \frac{1}{4}$.

   (c) Solve the equation in case $q = 1$.

   (d) Explain why the forms of the answers in the three cases above are different. (Hint: Look at the auxiliary equation.)

2. Show that $\cos(\ln x)$ and $\sin(\ln x)$ are solutions of $x^2y'' + xy' + y = 0$ and find the general solution.

3. Find the solution of the equation in (2) if $y(2) = 2$ and $y'(2) = 3$.

Laboratory Exercise 16.6

A Spring

Name ______________________________ **Due Date** ____________

An external periodic force is applied to a mass suspended by a spring. The mass moves according to the differential equation $y'' + y = 0.5\cos(\omega t)$. Assume initial values of $y(0) = 0$ and $y'(0) = 0$.

1. Explain in physical terms what the initial values say about the spring at time $t = 0$.

2. Solve the equation in the cases $\omega = 0.6,\ 0.8,\ 1, 1.2$, and 1.4 and plot all the graphs on the same screen.

3. The value $\omega = 1$ corresponds to the case that the period of the external force matches the natural period of the spring-mass. It produces a phenomenon known as *resonance.* Identify which graph this corresponds to. What do you expect will eventually happen to the spring in this case?

4. In the case $\omega = 0.6$, the motion of the spring-mass system can be seen from the graph to be periodic. Find this period. What is the total distance the mass moves over one period?

Appendix I

DERIVE Version 2.6 Reference

This appendix is a concise reference to the major features of *DERIVE* that often arise in the exercises in this book. Each topic refers the reader to the *DERIVE* User Manual (Version 2) by section for detailed information, and we encourage the interested reader to explore it.

As with any software, *DERIVE* evolves, and there are some features in versions 2.55 and later that were not present in earlier ones, such as the ability to print graphics. Two features are very prominent, and we warn users of earlier versions to be alert to them: (1) Plotting and (2) Line Editing. See Sections 2 and 6 for an explanation of the changes.

If you need help during a session, press [H] for **Help** and *DERIVE* will provide references to the *DERIVE* User Manual for the topic you choose.

Section 1. Moving Around

DERIVE Manual Section 3.3

When you have a long session with lots of expressions, it may become awkward to use the arrow keys to move up and down the screen. [Page Up] and [Page Down] move in bigger jumps. You can use [Home] and [End] to go between the top and bottom of the stack. To go to a given expression number, use the **Jump** command followed by the number of the expression.

Section 2. Line Editing

DERIVE Manual Section 2.4

Moving an expression to the author line: To change or edit an expression in the window, first highlight it by using the arrow keys [↑] [↓], press [A] for **Author**, and then press [F3]. You should see the expression appear on the author line where you can now edit it. Often you will want to bring an expression to the author line enclosed in parentheses. [F4] will do this.

Editing the expression: The left and right arrow keys can now be used to move the cursor on the author line *or* the highlight on the screen; [F6] toggles between these modes. Try it: Press the left and right arrow keys. Now press F6 and then press the left and right arrow keys again. In versions earlier than 2.55, the only way to move the cursor on the author line was with Ctrl S (left) and Ctrl D (right). This still works in version 2.55.

The Insert key toggles between typeover and insert modes; the Delete and Backspace keys erase. When you are finished, press Enter and the new expression appears on the bottom of the list of expressions.

Section 3. Splitting the Screen into Windows

DERIVE Manual Section 5.7

DERIVE versions 2.6 or later give you the option of splitting the window when you plot graphs, but with any version of *DERIVE*, you can split the screen into several windows to view multiple algebra and plot sessions at once.

Opening a new window: To split the active window into left and right pieces, use **Window Split Vertical**, or to split the window into upper and lower pieces, use **Window Split Horizontal**. You will be offered a line or column to split along. The default option splits the window in half.

Moving among open windows: When *DERIVE* has several windows open, it numbers them at the upper left corner and highlights the number of the active window. You can change the active window by pressing F1.

Designating a window type: There are three types of windows: 2D-plot, 3D-plot, and Algebra. You can change the type of the active window using **Window Designate**. When you change the type of a window, you lose all work or graphs that are currently there. When you change an Algebra window to a plot window, you will be prompted with Abandon expressions (Y/N). You must respond with Y to complete the type change.

Closing a window: To get rid of an unwanted window use **Window Close** and type the number of the window you wish to close at the prompt WINDOW CLOSE: Window: .

An example

1. **Author** x^2.
2. Issue the **Window Split Vertical** commands, then Enter, and you will see two windows numbered 1 and 2. Press the F1 key to flip between the windows.
3. To return to a full screen, highlight window 2 with the F1 key and issue the **Window Close** commands. Answer "Y" for "yes" to the prompt Abandon expressions (Y/N)?

Section 4. Defining Functions and Constants

DERIVE Manual Section 4.12

Defining functions: To define a function, for example $f(x) = 1 + x^2$, **Author** `F(x):=1+x^2` as seen in expression 1 of Figure 17.1. (Be sure to put the colon before the equal sign.) From now on *DERIVE* will know that F(x) is this function.

Evaluating defined functions: If you want to evaluate the function, $f(x)$, when $x = 3 + t$, **Author** `F(3+t)` and **Simplify**. The result appears in expression 3.

Redefining and undefining functions: If you want to redefine f, repeat the above process. You can even make f a generic function of x using `F(x):=`. The last definition you author is the one *DERIVE* accepts. If you no longer want f to be a function, **Author** `f:=` then Enter.

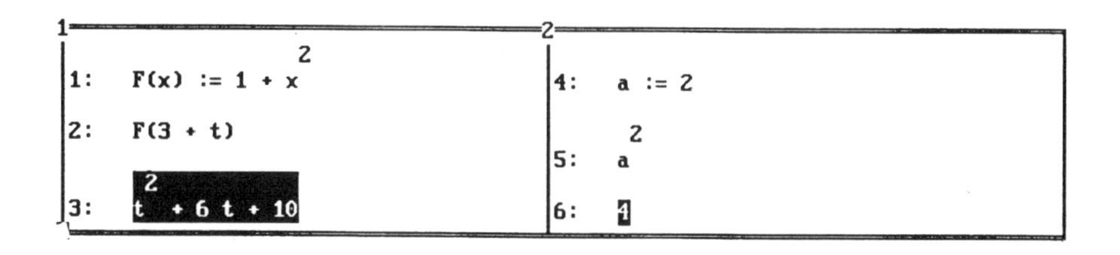

Figure 17.1: **Defining functions and constants**

Defining constants: To assign the value 2 to the letter a, **Author** `a:=2` as seen in expression 4 of Figure 17.1. When you **Author** `a^2` and **Simplify**, *DERIVE* reports the answer 4. If you want to re-define a, repeat the above process. The last definition you author is the one *DERIVE* accepts.

Undefining constants: If you no longer want a to be a specific constant, **Author** `a:=`.

Section 5. Substituting into an Expression

DERIVE Manual Section 4.8

The **Manage Substitute** option can be used to replace a variable or subexpression by an expression that is typed by the user, or by an expression that is already on the screen. We will

```
1:    x + COS(x + y) + SIN(x + y)

       3
2:    x

3:    x = y + 3

4:    5 + COS(5 + y) + SIN(5 + y)

       3         3             3
5:    x  + COS(x  + y) + SIN(x  + y)

6:    RHS(x = y + 3) + COS(RHS(x = y + 3) + y) + SIN(RHS(x = y + 3) + y)

7:    COS(2 y + 3) + SIN(2 y + 3) + y + 3

8:    x + COS(t) + SIN(t)
```

Figure 17.2: **Substituting into expressions**

illustrate the use of this option with an example. **Author** `x+COS(x+y)+SIN(x+y)`, `x^3`, and `x=y+3` as seen in expressions 1, 2, and 3 of Figure 17.2.

Replacing x: To replace x in expression 1 by the number 5, first highlight expression 1 and then use **Manage Substitute**. Press Enter to acknowledge the prompt MANAGE SUBSTITUTE expression: #1. Next type `5` at the prompt MANAGE SUBSTITUTE value:x. Finally, press Enter at the prompt MANAGE SUBSTITUTE value:y to indicate that the variable y is to be left unchanged. The result, with x replaced by 5, is in expression 4 of Figure 17.2.

Replacing x **by an existing expression**: To replace x in expression 1 by x^3 in expression 2, highlight expression 1 and use **Manage Substitute** as above. But at the prompt MANAGE SUBSTITUTE value:x, type `#2`. Press Enter for MANAGE SUBSTITUTE value:y, and the result appears in expression 5.

Replacing x **by the right-hand side of an existing expression**: To replace x by its value given by the equation in expression 3, we use the RHS (Right-Hand Side) function. Highlight expression 1 and use **Manage Substitute** as above, but at the prompt MANAGE SUBSTITUTE value:x, type `RHS(#3)`. **Simplify** the resulting expression 6 of Figure 17.2 to complete the substitution.

Replacing subexpressions *everywhere they occur in the formula*: To replace the subexpression $x + y$ everywhere it occurs in expression 1 by t, use the arrow keys to highlight any occurrence of $x + y$ that appears in the expression. (See expression 1 of Figure 17.2.) Now use **Manage Substitute**. Acknowledge MANAGE SUBSTITUTE expression:#1 with [Enter], and type `t` at the prompt MANAGE SUBSTITUTE value: . The result is in expression 8. Notice that both occurrences of $x + y$ have been replaced by t.

Section 6. Plotting Graphs

DERIVE Manual Section 5.2

To plot a graph, highlight the expression to be plotted and press [P] for **Plot**. If you do not already have a plot window open, you will be prompted for a choice: Beside Under Overlay. **Beside** opens a plot window to the right of your work, **Under** opens it below your work, and **Overlay** uses the full screen for plotting. (In versions prior to 2.55, the **Overlay** option is automatic, but Section 2 explains how to open a plot window manually.) If you use the **Beside** option, you will be prompted for a column number. The default is 40, but you may make it anything from 7 to 74. The screen will then split with an algebra window on the left and a plot window on the right.

You are now in the plot window. Press [P] for **Plot** again, and the highlighted expression will be graphed. You can use [A] for **Algebra** to get back to the calculations, and [P] for **Plot** to move back to the plot window. Alternatively, [F1] will toggle you between the two windows. (Look for the small numbers in the upper left corners.)

If you want to plot another expression, first be sure you have returned to the **Algebra** window, and then highlight the new expression you want to plot. If you now press [P] for **Plot** twice as before, *both* graphs will be plotted. To get rid of the original graph, use the **Delete First** command.

Plotting several graphs at once: If you want to plot three or more graphs, you can **Author** them in a list separated by commas and enclosed in *square* brackets. For example, to plot $\sin x$, $\sin(2x)$, and $\sin(3x)$, **Author** `[SINx, SIN(2x), SIN(3x)]` as seen in window 1 of Figure 17.3. When you **Plot** this list, all three graphs will be drawn in order. (See window 2.) If you want just two graphs, for example $\sin x$ and $\sin(2x)$, you may **Author** and **Plot** `[SINx, SIN(2x), ?]`. If you omit the question mark and just **Plot** `[SINx, SIN(2x)]`, *DERIVE* will interpret it as a single parametric curve.

Plotting a graph over a specific interval: To plot x^2 over the interval $[-1, 0.5]$, **Author** `[x, x^2]` as seen in window 3 of Figure 17.3. Now press [P] for **Plot** and [Enter] [Enter] to

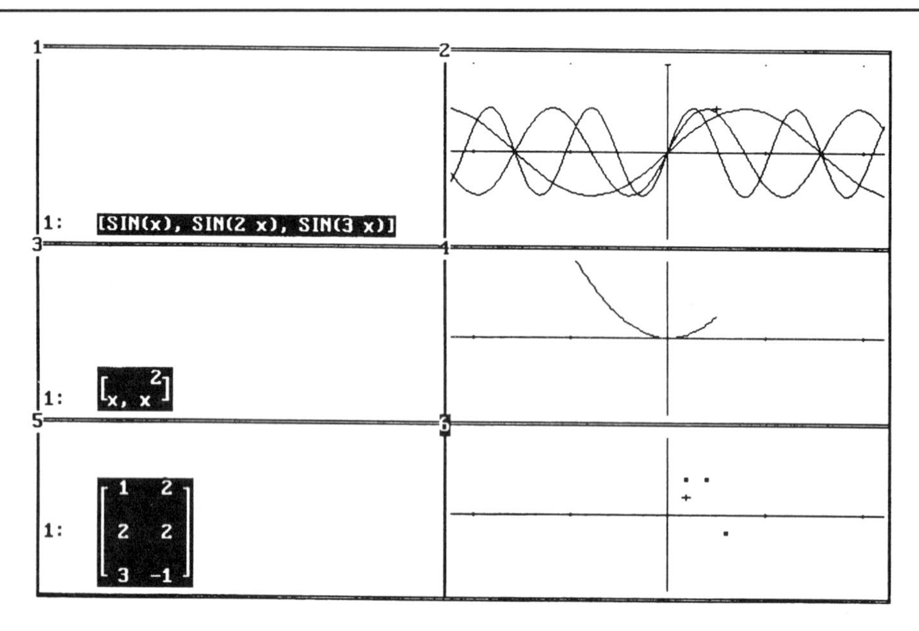

Figure 17.3: **Multiple graphs, restricted domains, and discrete plots**

choose **Beside** 40. You will see the prompt PLOT: Min:-3.1416 Max:3.1416 at the bottom of the screen. This is where you set the interval. In the Min: field, Delete -3.1416 and type `-1`. Use the Tab key to move to the Max: field, Delete what's there, and type `0.5`. Now press Enter and the graph will be drawn. (See window 4 of Figure 17.3.)

Plotting individual points. Suppose you want to plot several points on the screen, for example, (1, 2), (2, 2), and (3, −1). **Author** the list `[[1,2],[2,2],[3,-1]]`. It appears as a *matrix* in window 5 of Figure 17.3. When we **Plot** it, the individual points are shown in window 6.

Plotting in polar coordinates: To plot the polar equation $r = \sin(3\theta)$, **Author** the expression `SIN(3`Alt H`)` as seen in window 1 of Figure 17.4. (Alt H or `theta` makes the Greek letter θ. You can use some other letter if you like.) Now **Plot**.

When you are in the plot window, use **Options State Polar** Enter to change *DERIVE's* graphics settings from Rectangular to Polar. (You only need to do this once. The state will remain the same until you change it back.) Now issue the **Plot** command once more. *DERIVE* will offer us a "parameter domain" (a range for θ) that we set to Min:`-pi` Max:`pi`. (See **Plotting over a specific interval** above.) We see a "three-leafed rose" in window 2 of Figure 17.4.

Plotting parametric curves: To plot the graph of the parametric curve, $x = \sin t$, $y = \sin(2t)$

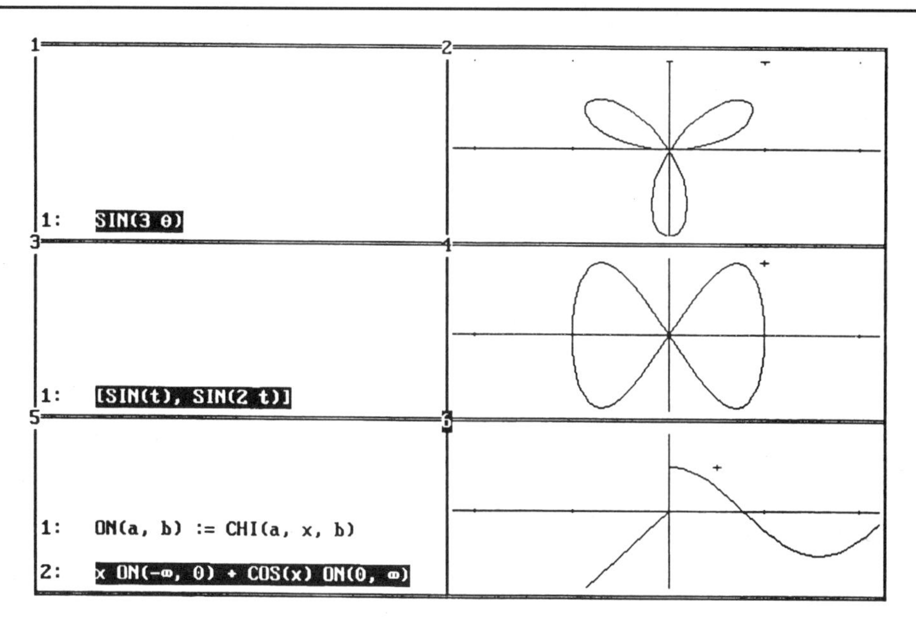

Figure 17.4: **Polar, parametric, and piecewise plots**

on the interval $[-\pi, \pi]$, we first **Author** `[SINt,SIN(2t)]` as seen in window 3 of Figure 17.4, and then **Plot**. (If you've been doing polar plots, be sure to change **Options State** in the plot window to **Rectangular**.) **Plot** again, and set the "parameter domain" to Plot Min:`-pi` Max:`pi`. (See **Plotting over a specific interval** above.) The graph is in window 3 of Figure 17.4.

Plotting piecewise defined functions: You can accomplish this using *DERIVE's* CHI function. CHI(a, x, b) is equal to 1 on the interval (a, b) and 0 elsewhere. You may find it helpful to rename this function by **Author**ing `ON(a,b):=CHI(a,x,b)` as seen in expression 1 of window 5 of Figure 17.4. This provides a mnemonic syntax for defining piecewise functions. For example, if we want x on $(\infty, 0)$ and $\cos x$ on $(0, \infty)$, we can **Author** `x ON(-inf,0)+ (COS x)ON(0,inf)`. **Plot** to get the graph in window 6.

Warning: The IF function is intended for use as a logical construct and may cause problems if you use it to define a function piecewise. For example, if you **Author** and **Plot** `IF(x<0, x^2, x+1)`, it gives the same graph as `x^2 ON(-inf,0)+ (x+1)ON(0,inf)`, but try using **Calculus Limit** to find the limit of it as $x \to 0$ from the right and from the left. Do the same using the definition `x^2 ON(-inf,0)+ (x+1)ON(0,inf)` and compare.

Summary of Plot Features

- **Plot** moves from the algebra window to the plot window.
- **Algebra** returns to the algebra screen from the plot screen.
- [Esc] stops plotting. [P] for **Plot** starts it.
- The four arrow keys move the small cross on the screen. Notice that the coordinates of the cross appear in the lower left corner. [Ctrl ←], [Ctrl →], [Page Up], and [Page Down] move the cross in bigger jumps.
- **Center** centers the graph on the cross.
- **Scale** lets you change the scales on the x and y axes.
- [F9] zooms in, and [F10] zooms out.

Five Common Questions about Plotting

The graph is just a bunch of little dots. What's wrong? You are not in graphics mode. Use the **Options Display** commands in the plot window to change from Text to Graphics and also to set the correct graphics adapter. (See the *DERIVE* Manual Section 5.1.) After this, [F5] will flip between text and graphics modes if required. (You may save your two most recent screen modes for later sessions with the **Transfer Save State** commands.)

I tried to plot a new graph, and I got another one that I did earlier. How do I get rid of it? *DERIVE* saves all your 2D-plots. If you plot a graph, then go to the algebra screen and plot another; the first will be graphed followed by the second. Use the command **Delete**, and you see the prompt: DELETE All Butlast First Last. These four choices are self-explanatory, but if you just want the graph you last asked to plot, press [A] for **All** and then [P] for **Plot**.

I tried to plot something, and DERIVE beeped at me. In the lower left corner of the plot screen I saw the message "Cannot do implicit plots." What's wrong? You probably are trying to plot an equation such as $x^2 + y^2 = 1$ or $y^2 = x^2$. *DERIVE* plots *expressions* but not equations. There is an exception: *DERIVE* will plot an equation of the form $y = f(x)$ as if it were $f(x)$ alone. To plot an equation such as $x^2 + y^2 = 1$, you may solve for y and plot the two solutions $\pm\sqrt{1 - x^2}$ together.

Note: At the time of printing of this book, the release of *DERIVE* 3.0 is imminent. Version 3.0 *will* make implicit plots.

How do I plot a vertical line? To plot the line $x = 3$, for example, **Author** `[3,y]` and **Plot** it. This is a parametric plot, and *DERIVE* will ask for a parameter interval. If you set this to Min:`-1` Max:`2` you will get a plot of the line $x = 3$ from $y = -1$ to $y = 2$. (See **Plotting a graph over a specific interval** above.)

I've gotten my plot window completely messed up. How do I reset it to the original default configuration? Use **Window Designate 2D** (or **Window Designate 3D** for a 3D-plot window).

Section 7. Solving Equations Exactly

DERIVE Manual Section 4.14

Solving equations: To get the exact solution of an equation, **Author** it and use the **soLve** command. If the precision is set on **Exact** (see the next section on *Approximations and Precision*), *DERIVE* will attempt to give the exact roots.

Author and **soLve** the equation `2x^4 + 9x^3 - 27x^2 = 22x - 48`. The solutions appear in expressions 2 through 5 of Figure 17.5.

Solving expressions: If you ask *DERIVE* to **soLve** an expression that is not an equation, inequality, or other relation, it will try to find the zeros of the expression. Try it for $x^2 + 4$. Notice in expressions 7 and 8 that *DERIVE* works with complex numbers.

Equations that *DERIVE* cannot solve: When a solution cannot be found exactly, *DERIVE* simply returns the equation (often with some adjustments). This has occurred for $x^5 + 5x^2 - 1$ in expressions 9 and 10. In this case it will be necessary to approximate the solutions. See Section 9.

Equations that have no solutions: If *DERIVE* is able to determine that there are no solutions, it will beep and say "No solutions found." In expression 11 of Figure 17.5 we have attempted to solve $x + 1 = x - 1$. Note the message at the bottom of the figure.

Section 8. Approximations and Precision

DERIVE Manual Section 3.8

Changing between exact and approximate modes: If you want *DERIVE* to *always* give approximate answers, choose **Options Precision Approximate**. To make it give exact answers choose **Options Precision Exact**. Don't forget to reset this switch at the appropriate time. When *DERIVE* is set on **Exact**, you may still get decimal approximations to numbers using [X] for **approX**.

```
2                                        3
         4       3        2
1:   2 x  + 9 x  - 27 x  = 22 x - 48
                                           2
          3                        6:   x  + 4
2:   x = - ─
          2                        7:   x = - 2 î

3:   x = 2                         8:   x = 2 î

         √57    5                       5      2
4:   x = - ─── - ─                 9:   x  + 5 x  - 1
          2     2
                                        5      2
         √57    5                  10:  x  + 5 x  = 1
5:   x = ─── - ─
          2     2                  11:  x + 1 = x - 1
COMMAND: Author Build Calculus Declare Expand Factor Help Jump soLve Manage
         Options Plot Quit Remove Simplify Transfer moVe Window approX
No solutions found
User                                   Free:99%                Derive Algebra
```

Figure 17.5: **Solving equations exactly**

Setting precision: To set the number of decimal places that *DERIVE* uses for approximations, use the **Options Precision** commands. The Tab key jumps to the number where you indicate how many decimal places you would like.

Section 9. Solving Equations Approximately

DERIVE Manual Section 4.14

We found in Section 7 that *DERIVE* cannot solve the equation $x^5 + 5x^2 = 1$ exactly. If *DERIVE* is set to its approximate mode (see Section 8), and you ask it to **soLve**, it will apply a technique called *the bisection method* to find an approximate answer. To do this, it must begin with an interval in which to search for a root. The procedure for getting all the solutions is outlined below.

Step 1: Write the equation in the form *expression* $= 0$ **and Author the left-hand side.** In this case, we **Author** `x^5+5x^2-1` as seen in expression 1 of Figure 17.6.

Step 2: Plot the graph and locate search intervals: The graph appears in window 2 of Figure 17.6, and we see that there is one zero on each of the intervals $[-2, -1]$, $[-1, 0]$, and $[0, 1]$.

Step 3: Change to the approximate mode: Use **Algebra** to get back to the calculations, and then use **Options Precision Approximate**.

Step 4: soLve on each of the search intervals found in Step 2. With expression 1 of Figure 17.6 highlighted, **soLve** setting the Lower: field to -2 and the Upper: field to -1. (Use the Tab key to set the Upper: field.) The first solution appears in expression 2 of Figure 17.6. Highlight expression 1 again and **soLve** on the range Lower:-1 Upper:0 to get the solution in expression 3. **soLve** once more on the range Lower:0 Upper:1.

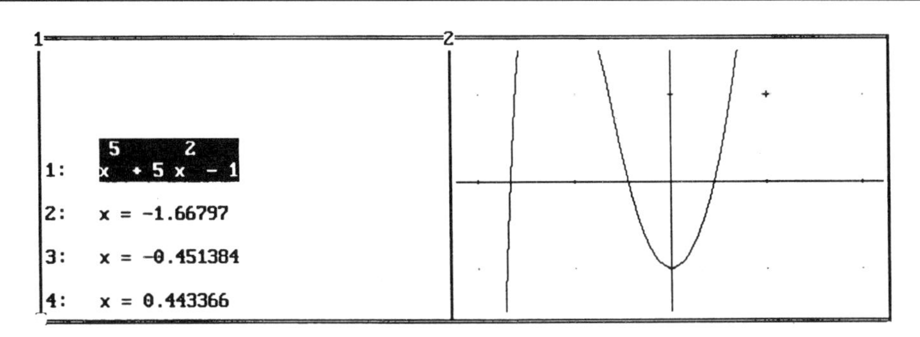

Figure 17.6: **Solving in the approximate mode**

In summary, the bisection method employed by *DERIVE* will find at most one root in the prescribed interval no matter how many there may actually be. Therefore, it is a good idea to plot the graph to determine the search intervals. In some cases speed may be an important factor: The smaller the interval, the quicker *DERIVE* is able to get the answer.

A Strategy for Solving Equations

Step 1: Ask *DERIVE* to **soLve** in the exact mode . If all the solutions are found, you do not need to proceed further.

Step 2: If exact solutions are not found, write the equation as $f(x) = 0$, **Plot** $f(x)$, and isolate the zeros in small intervals.

Step 3: Set *DERIVE* to its approximate mode and **soLve** $f(x)$ on each of the intervals found in Step 2.

Step 4: Be alert to the possibility that zeros may be hiding outside the range of the plot. You may zoom out with F10 to convince yourself that there are no more, but in general, some mathematics may be required to establish that all zeros have been found.

Section 10. Calculus

DERIVE Manual Chapter 7

The calculus operations are under the **Calculus** command. The options include **Differentiate**, **Integrate**, **Limit**, **Product**, **Sum**, and **Taylor** (Taylor polynomials).

Integration: To Integrate x^2+3x-4, **Author** `x^2 + 3x - 4` as seen in expression 1 of Figure 17.7 and select **Calculus Integrate**. Press Enter twice to accept the defaults, expression: #1 and variable: x. For a definite integral, we would type the limits of integration at the prompt, Lower limit: Upper limit: . (The Tab key jumps between the two.) To get the indefinite integral, just press Enter with no limits. The integral appears in expression 2. To evaluate the integral, use **Simplify** and we see expression 3 of Figure 17.7.

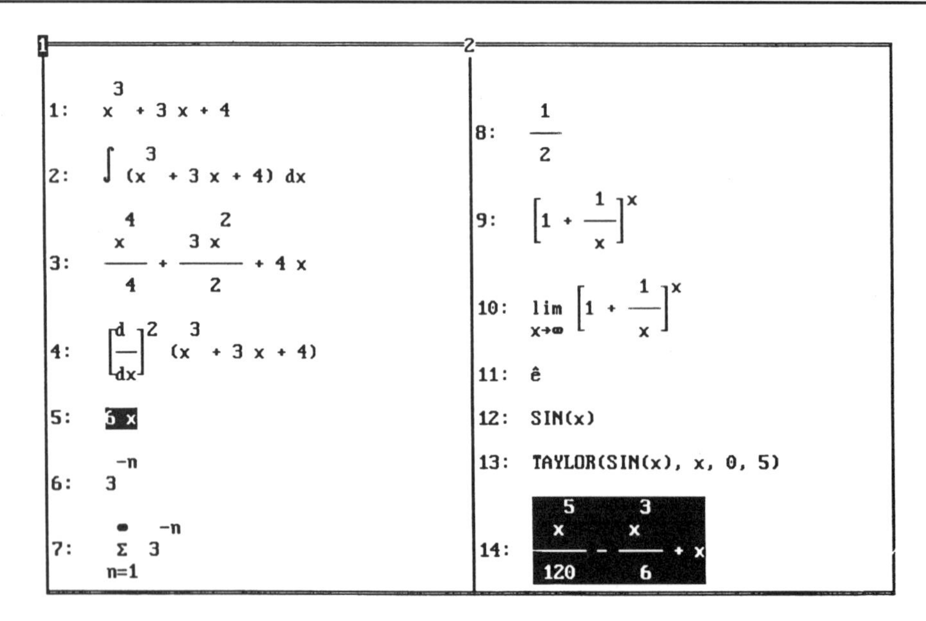

Figure 17.7: **Calculus operations**

Derivatives: Find the second derivative of $x^2 + 3x - 4$. Highlight expression 1 of Figure 17.7, and select **Calculus Differentiate**. Press [Enter] twice to accept the defaults, expression: #1 and variable: x. Now we see the prompt CALCULUS DIFFERENTIATE: Order: 1. We want order two, so type 2 and then [Enter]. We see *DERIVE's* notation for the second derivative in expression 4. Most calculus books use $\frac{d^2}{dx^2}(x^2 + 3x - 4)$. To evaluate it, use **Simplify**, and we see the answer in expression 5.

Sums: To find the sum of the series $\sum_{n=1}^{\infty} 3^{-n}$, **Author** `3^-n` as seen in expression 6 of Figure 17.7. Use **Calculus Sum** and [Enter] [Enter] to accept the defaults, expression: #6 and variable:n. Next we see the prompt CALCULUS SUM: Lower limit:1 Upper limit:n. Use the [Tab] key to move to the Upper limit: field, and type `inf` in place of n, and then [Enter]. The sum appears in expression 7 of Figure 17.7. When you **Simplify** it, you get the answer, $\frac{1}{2}$, in expression 8.

Limits: To find the limit as x goes to infinity of $\left(1 + \frac{1}{x}\right)^x$, **Author** `(1 + 1/x)^x` as seen in expression 9 of Figure 17.7. Use **Calculus Limit** and accept the first two defaults as in the examples above. When prompted for Point: type `inf` and [Enter]. The limit appears in expression 10. **Simplify** to get the answer, e, in expression 11.

Taylor Polynomials: To find the Taylor polynomial of degree 5 for $\sin x$, **Author** `SINx` and use **Calculus Taylor**. Accept the defaults, expression:#12 and variable:x, by pressing [Enter] twice. You see the prompt Degree:5 Point:0. These can be changed to accommodate any degree and any expansion point, but these are what we want, so we press [Enter] once more. The notation for the Taylor polynomial appears in expression 13 of Figure 17.7. **Simplify** to get the answer in expression 14.

Appendix II

Optional Files for Producing Graphics

The code in this appendix is designed to produce graphical displays of important concepts from calculus. Be careful to **Author** the expressions exactly as they appear. Take special care to distinguish square brackets from parentheses, don't omit any commas, and be sure to use ":=" instead of just "=".

Riemann Sums: The RIEMANN.MTH File

This code is designed to produce graphical displays of left-hand, right-hand, or midpoint Riemann sums and of the trapezoidal rule. Begin with a clear screen and be careful to **Author** the expressions exactly as they appear below.

```
F(x):=

(b-a)/n

[[p,0],[p,F(p+z#2)],[p+#2,F(p+w#2)],[p+#2,0]]

VECTOR(#3,p,a,b-#2,#2)

PIC(a,b,n,z,w):=[F(x),#4]

LPIC(a,b,n):=PIC(a,b,n,0,0)

RPIC(a,b,n):=PIC(a,b,n,1,1)

MPIC(a,b,n):=PIC(a,b,n,1/2,1/2)

TPIC(a,b,n):=PIC(a,b,n,0,1)
```

Save this as a file for future use with **Transfer Save** RIEMANN. When you want to recall it, use **Transfer Merge** RIEMANN.

Instructions for using RIEMANN.MTH file: The function LPIC(a, b, n) prepares *DERIVE* to produce a picture of the left-hand Riemann sum for the function $F(x)$ on the interval $[a, b]$ with n subintervals. Similarly, RPIC makes a picture of right-hand sums, MPIC makes a picture of midpoint sums, and TPIC makes a picture of the trapezoidal rule. Special graphics settings are required to make the pictures. See the example that follows.

Example: We will produce a picture of the midpoint sum for $f(x) = x^2$ on $[0, 1]$ with $n = 10$ subintervals.

Step 1: **Author** `F(x):=x^2` to tell *DERIVE* which function we want.

Step 2: **Author** and **approX** `MPIC(0,1,10)`.

Step 3: **Plot** to open a plot window, and then change the graphics settings using **Options State Rectangular Connected Small**.

Step 4: **Plot** again to make the picture.

Simpson's Rule: The SIMP.MTH File

Simpson's rule has a very interesting graphical interpretation as the integral of quadratic approximations to the graph (parabolas). This code is designed to produce this graphical display of Simpson's rule. Begin with a clear screen, and be careful to **Author** the expressions exactly as they appear below.

```
F(x):=

(b-a)/n

VECTOR([i,F(i)t],i,a,b,#2)

t(2t-1)F(i+2#2)

4t(1-t)F(i+#2)

(t-1)(2t-1)F(i)

[i+2t#2, #4+#5+#6]

VECTOR(#7,i,a,b-2#2,2#2)

SPIC(a, b, n) := [#3,#8]
```

Save this as a file for future use with **Transfer Save** SIMP. When you want to recall it, use **Transfer Merge** SIMP.

Instructions for using the SIMP.MTH file: The function SPIC(a, b, n) prepares *DERIVE* to produce a picture of Simpson's rule for the function $F(x)$ on the interval $[a, b]$ with n subintervals. (n *must be even*; otherwise, the picture will not represent the approximation correctly.) The plot is *parametric*, and the parameter domain *must* be set to $[0,1]$. See the example that follows.

Example: We will make a picture of Simpson's rule for $F(x) = \sin x$ on the interval $[-2\pi, 2\pi]$ using $n = 6$ subintervals.

Step 1: **Author** `F(x):=SINx` to tell *DERIVE* which function we want.

Step 2: **Author** and **approX** `SPIC(-2pi, 2pi, 6)`.

Step 3: **Plot** to open a plot window, **Plot** again, and set PLOT: Min:0 Max:1.

Step 4: Use Ctrl Enter to get the picture.

Newton's Method: The NEWTON.MTH File

This code is designed to produce a graphical display of Newton's method. Begin with a clear screen, and be careful to **Author** each line exactly as it appears below.

```
F(x):=

x-F(x)/DIF(F(x),x)

ITERATES(#2,x,a,n)

ELEMENT(#3,i)

ELEMENT(#3,i+1)

[#4,F(#4)t]

[(1-t)#4+t#5,(1-t)F(#4)]

[#6,#7]

NPIC(a,n):=[F(x), vector(#8,i,1,n)]
```

Save this as a file for future use with **Transfer Save** NEWTON. You can call it back when you like using **Transfer Merge** NEWTON.

Instructions for using the NEWTON.MTH file: The function NPIC(a, n) prepares *DERIVE* to make a picture of n iterations of Newton's method for the function $F(x)$ beginning at the point $x = a$. The plot is *parametric*, and the parameter domain *must* be set to $[0,1]$. See the example that follows.

Example: We will show four iterations of Newton's method applied to $F(x) = x^2 - 2$ with starting point $x_0 = 3$.

Step 1: **Author** `F(x):=x^2-2` to tell *DERIVE* which function we want.

Step 2: **Author** and **approX** `NPIC(3, 4)`.

Step 3: **Plot** to open a plot window, **Plot** again, and set PLOT: Min:0 Max:1.

Step 4: Use Ctrl Enter to get the picture.

Increasing and Decreasing Functions: The INCREASE.MTH File

On a color monitor, this file plots the graph of $f(x)$ in one color where it is increasing and in another color where it is decreasing. Begin with a clear screen, and be careful to **Author** the expressions exactly as they appear below.

```
F(x):=

DIF(F(x),x)

IF(#2>0, F(x))

IF(#2<0, F(x))

plot_this:=[#3, #4, ?]
```

Save this as a file for future use with **Transfer Save** INCREASE. You can call it back when you like using **Transfer Merge** INCREASE.

Instructions for using the INCREASE.MTH file: After you tell *DERIVE* which function you want, **approX** the line beginning with `plot_this` to make the picture.
Example: We will plot the graph of $\sin x$ in one color where it is increasing and in another where it is decreasing.

Step 1: **Author** `F(x):=SINx` to tell *DERIVE* which function we want.

Step 2: Highlight the line beginning with `plot_this` and **approX**.

Step 3: **Plot** to open a plot window, and then **Plot** again.

Concavity: The CONCAVE.MTH File

On a color monitor, this file plots the graph of $f(x)$ in one color where it is concave up and in another color where it is concave down. Begin with a clear screen, and be careful to **Author** the expressions exactly as they appear below.

```
F(x):=

DIF(F(x),x,2)

IF(#2>0, F(x))

IF(#2<0, F(x))

plot_this:=[#3, #4, ?]
```

Save this as a file for future use with **Transfer Save** CONCAVE. You can call it back when you like using **Transfer Merge** CONCAVE.

Instructions for using the CONCAVE.MTH file: After you tell *DERIVE* which function you want, **approX** the line beginning with `plot_this` to make the picture.

Example: We will plot the graph of $\sin x$ in one color where it is concave up and in another where it is concave down.

Step 1: **Author** `F(x):=SINx` to tell *DERIVE* which function we want.

Step 2: Highlight the line beginning with `plot_this` and **approX**.

Step 3: **Plot** to open a plot window, and then **Plot** again.

Secant Lines: The SECANT.MTH File

This file makes a picture of secant lines to the graph of f.

```
F(x):=

(F(a+ks)-F(a))/(ks)

(x-a)#2 + F(a)

VECTOR(#3, k, n, 1, -1)

SECANT(a, n, s):=[F(x), #4]
```

Save this as a file for future use with **Transfer Save** SECANT. You can call it back when you like using **Transfer Merge** SECANT.

Instructions for using the SECANT.MTH file: The function $\text{SECANT}(a, n, s)$ makes a picture of n secant lines through the points $(a, f(a))$ and $(a + ks, f(a + ks))$ as k ranges from 1 to n in steps of s.

Example: We will plot five secant lines to the graph of $\sin x$ at $a = 1$ using a step size of $s = 0.2$.

Step 1: **Author** `F(x):=SINx` to tell *DERIVE* which function we want.

Step 2: **Author** and **approX** `SECANT(1, 5, 0.2)`.

Step 3: **Plot** to open a plot window, and then **Plot** again.

Index of Solved Problems

Index of Laboratory Exercises